In
Defense
Of My Country

Indo-US Defense Technology Cooperation

Issues and Opportunities

Rayol John Augustus Ph.D.

R. J. Augustus

In Defense of My Country
By Rayol John Augustus

Copyright © 1998 Rayol John Augustus
All Rights Reserved.

ISBN-13: 978-1975919962
ISBN-10: 1975919963

Cover Design : Prof. Meena Augustus Ph.D.

Contents

Preface	xi
Acknowledgements	xiv
1 Introduction	1
2 The New International System	9
Changes In The Military threat	13
Why Proliferation ?	15
Post-Cold-War Reality	18
Third World's Demands For Weaponry	19
New Technology - The Need To Modernize	21
Arms Race On The Subcontinent	23
3 An Anatomy of Indo US Relations	29
Interdependence Among Nations	37
Role Of The US And India	39
Why India Matters	42
The Defense Technology Scene	44
4 The Export Control Problem	49
Arms Transfers	50
Appeal Of Arms Exports To Governments	52
The Arms Exporting Company	58

Arms Importing Governments	62
Influence of Exports On Foreign Policy	64
Decline Of The Global Arms Market	66

5 Indian Perspectives 71

Indian Neighborhood	72
The China Factor	74
The Pakistan Factor	80
Indian Security Strategy	82
Regional Relations	85
Defense Spending Among South Asian Neighbors	87

6 The United States' Perspectives 95

US Strategy In The Asia-Pacific	99
US Technology Strategy	101

7 Arms Control 105

Conventional Arms	105
Non-Proliferation Regimes	110
Missile Technology Control Regime	110
MTCR - Strengths And Weaknesses	113
Non-Proliferation Treaty	114
Comprehensive Test Ban Treaty	117
Wassenaar Arrangement	121
The Groups	123
India's Approach To The Regimes	125

8 India-US High Technology Transfer 133

Technology Transfer	136

US Congress Legislation on Technology Transfer	141
DOD Policy On Technology Transfer	143
The Technology Transfer Process	145
Technology Transfer Mechanisms	152
The Trade - Off	156
Articles of Agreement	158
Technology Transfer Evaluation	159
Technology Partnerships	160
Spin-offs	162
Indian Acquisition Cycle	162
Barriers To Technology Transfer	167
The Export License	171

9 Indian R&D Scenario — 177

Defining New And Emerging Technology	177
Critical, Core And Dual Use Technologies	179
India's March To Self-Reliance	183
India's R&D Strategy	189
India's Missile Thrust	193
The Right To Affordable Technology	195
Technology Absorption And Management	196
Industrial Cooperation	200
DRDO And Societal Goals	203

10 Export Controls In A Changing World — 205

The Trouble With Arms Exports	208
US Arms Sales	212
US Industry-Government Interaction	217
Commercial Or Foreign Military Sales	219

11 Indian Participation In Controls		223
Indian Import Controls		223
Export Controls In India		225
12 Strengthening Indo-US Relations		231
Indo-US Defense Security Cooperation		233
13 Conclusion		243
India - A Time to Change		245
US – A Time To Rethink Policy		255
Finding Common Ground		259
Addendum		269
Appendices		271
References		283
Glossary of Acronyms		307
Index		313
About the Author		325

In Defense Of My Country

List of Illustrations

No.	Figure	
1	US Relations With India	30
2.	Leading Military Spenders	68
3.	Indian Defense Expenditures	88
4.	Military Expenditure Of India, Pakistan and China	92
5.	Import of Major Conventional Weapons	92
6.	Indian Defense Trends	93
7.	Technology Transfer Features	137
8.	Technology Transfer Cycle	138
9.	Technology Transfer Drivers	140
10.	Technology Transfer Process	145
11.	Technology Transfer Environment	148
12.	Technology Transfer Participants	148
13.	Technology Transfer Indicators	149
14.	Technology Transfer Mechanisms	151
15.	Technology Availability & Trade-Off	153
16.	Technology Transfer Articles of Agreement	154
17.	Technology Acquisition / Procurement Cycle	159
18.	Indian Tech Transfer & Absorption–DRDO to DP	161
19.	Barriers To Technology Transfer	166
20.	Indian R&D Strategy Today	186
21.	Technology Transfer / Absorption Practice	187
22.	Transformation Of Indian Industry	188
23.	US Foreign Military Sales Agreements	207
24.	US Foreign Military Sales Deliveries	208

25. US-India Foreign Military Sales	214
26. Indian Import Certificate	219
27. Indo-US MOU-IC Procedures For GOI	262
28. Indo-US MOU-IC Procedures For Private Sector	264

List of Tables

Table No   **Table**

1. Indian Expenditure On Military R&D   86
2. Indian Tax Revenue & Expenditure   87

List of Appendices

Appendix		Page
1.	Form for to be submitted by the applicant to the Import Certificate Issuing Authority For Issue of the Import Certificate.	259
2.	Form to be submitted by the applicant to the Import Certificate Issuing Authority For Higher Assurances.	261
3.	Steps in the Import Certificate Procedures for Government to Government - Controlled Entities	262
4.	Steps in the Import Certificate Procedures for Private Sector Entities	264
5.	Delivery Verification Procedures	266

Note: The Forms and Steps presented in the Appendices are guidelines and may change from time to time.

 # Preface

It was Freeman Dyson, British born physicist who said ' If we had a reliable way to label our toys good and bad, it would be easy to regulate technology wisely. We can rarely see far ahead to know which road leads to damnation. Whoever concerns himself with big technology, either to push it forward or to stop it is gambling in human lives.' Technology is not a separate entity that is remote from the lives of people, for it provides the basis for most of the requirements of modern living. Though defense technology has for some time been identified singularly with military applications, today every bit of it can be considered 'dual use' with an equal number or more of civilian applications. The area of International Technology Transfer should be entitled to a special place in diplomatic relations. Technology transfer brings people together. It brings countries together. To make a success of a technical project, people learn to live with each other, respect each other's points of view, work together and navigate through barriers and strive to make it a win-win situation which is what diplomatic relations is all about. All too often technology transfer is used as a punching bag for diplomatic sparring.

My own study tries to navigate in the troubled waters of defense technology transfer, export controls and issues and opportunities for a developing country. The world is no longer the same and old thinking cannot deal with today's issues. The drive for power by men and nations traditionally has been checked by the drive for power of other men and nations. This combined with fear of an unstable world forces some nations to adopt measures that may not be acceptable to all. Defense Technology is a victim of these circumstances and it is ironical that those nations that want to control it the most are the ones who want to export it the most.

India's search for security faces the constraints of her location, size, population and historic pattern of conflict and alliance peppered with the politics and economics of a changing world scenario. Reduction in threat perceptions followed by technological, military and economic revolutions require nations to rethink their political and defense strategies. India and the US are no exceptions. Each nation has its own strategic plan which at times presents hurdles in their relationship. America, the greatest world power, is beginning to lose its grip on technological leadership to other runners. India, on the other hand, with a growing economy and technological capability, is emerging as a world power and has to be considered seriously. India's defense strategy calls for force modernization and technology self-reliance. The technology and transfer required, from the US and other western countries, to build this capability, is inhibited by western strategy of export controls and non-proliferation regimes that third world countries consider discriminatory.

I am no writer. This book is a transformation of my 1996 doctoral dissertation "Technology Transfer To Indian Defense R&D laboratories Through Indo-US Defense Cooperation- Issues And Opportunities", revised up until early May 1998. My only desire is for it to escape the conventional dusky corner of a library that all dissertations seem to gravitate towards. It analyzes Indo-US relations and defense technology cooperation, keeping in mind regional situations and brings out the perspectives, fears and reactions of India and America to global concerns. A study has been made of various factors influencing technology transfer and transfer models have been developed with reference to India's R&D strategy. This study emphasizes the necessity for both countries to explore opportunities in finding working equations albeit their many differences on many issues.

An attempt has been made to understand concerns of both countries and to advocate ways in which Indo-US bilateral defense technology transfer can be improved through confidence building and restoring measures. Recommendations are made on how Indian R&D can be improved for better acquisition and absorption of transferred technology. A positive Indian approach and change in American export control policy are projected so that India and America can find common ground in their technological interaction through the years ahead.

I have tried to examine the international situation today, the technological market, the appeal for the export and import of defense technology, the technology transfer problem, the Indian R&D environment, the control regimes and Indo-US defense cooperation with the basic assumption that both India and the US have to make repeated efforts to work together to find acceptable solutions. The views, analytical approaches and conclusions are based on definite values and theoretical premises with the belief that the issues can be tackled only with transparency and cooperation rather than animosity. It should be a cooperation on the basis of equality, mutual understanding, respect and reciprocity and this should be done in all urgency before the opportunities that we have today are lost to history.

Washington DC *Rayol John Augustus*
May 1998

R. J. Augustus

Acknowledgements

The magnitude of the Indo-US technology transfer issue dawned upon me on my posting to the Embassy of India in Washington D.C. as Adviser in Defense Technology. In that office, a key concern was to interact with the US Government and US industrial houses on technology and technology transfer. If there is anything that one learns while working in Washington D.C. is that there is almost no time for reflection. Working the problem daily, extensive study during stolen weekends and holidays, interesting readings while waiting at airports and staying at hotels during temporary duty moves all over the United States have all contributed to this book. This work has been a personal challenge, but worth every bit of the effort as it has broadened my horizons and given me a deep personal satisfaction.

Two major opportunities in my professional career are due to Dr. K.G. Narayanan, Distinguished Scientist and Director, Aeronautical Development Establishment, DRDO, Ministry of Defense, Government of India. How do I thank him ?

When I was posted to Washington D.C., Mr. K. Santhanam, Chief Adviser (Technology), DRDO, suggested that I do some concentrated study during my tenure in the USA. I must thank him for starting me on this path and his continued support. Thanks are due to Dr. A.P.J. Abdul Kalam, Scientific Adviser to the Defense Minister, Government of India, who urged me to work on Technology Transfer and the issues involved.

In Defense Of My Country

I am deeply grateful to Dr. V.S. Arunachalam, former Scientific Adviser to the Defense Minister, Government of India, now Professor at the Carnegie Mellon University, Pittsburg, for all the useful leads he gave me when I strayed from my present subject.

No scholar can operate without institutional help and the Defence Research and Development Organization, of which I have been a part for the last twenty five years, deserves my sincere thanks. Thanks are due to scientists I know intimately in this organization and those whom I never met who helped me and influenced me indirectly. To my colleagues at the Embassy of India, I am grateful for the intellectual stimulation our interaction has given me. I must add that all opinions and ideas are mine alone and do not represent in anyway those of the DRDO or the Government of India.

If at times there appear traces of any sort of national bias in the arguments, I seek refuge under the shelter of human frailty. Needless to say that all errors, generalizations, biases and flaws are mine. My family is responsible for the better part of this book. This book would not have been possible without the encouragement and support of my wife, Prof. Meena Augustus. To her I am deeply grateful. To my mother and children I am most thankful for their kindness and understanding when I stole away all those weekends and quiet evening hours.

Rayol John Augustus

1
Introduction

Like the coconut palms that fringe its coast line, India too is resilient. This has been proved every time India came through trauma after trauma, whether it was a natural calamity, a military invasion or internal political turmoil. When one views the subcontinent from way above, beyond manmade boundaries, it is easy to see why geography has profoundly affected India's history and culture and her strategic thinking. The Himalayas in the north, the Indian Ocean in the south, the Bay of Bengal in the east, and the Arabian Sea in the West have created a giant natural entity referred to as the Indian subcontinent. The subcontinent stretches over 3000 kilometers from the north to south and roughly the same distance from east to west. India, the largest part of the subcontinent is spread over 3 million square kilometers and populated with over 960 million people, making it the second largest nation in the world. It is easy to jump to the conclusion that India's strategic location, size and vast population must make her great and with international clout. However because of this false belief, India has had to play a secondary role over the years. Today, however Indians have come to realize that geographical settings, size, population and cultural heritage alone do not matter but the collective will of the people to give better than their best and to compete at a global level instead of fighting among themselves.

R. J. Augustus

Every nation's search for security faces certain constraints of geographical location, size, population and historic pattern of conflict or alliance peppered with the strong influence of politics which is in a constant state of flux. India and the US are no exceptions and each nation has its own strategic plan which at times presents hurdles in the defense relationship. Yet, the United States defense industry increasingly sees India as an important market for US defense technology with its associated products and services. India, with its large defense force structure and inventory, necessary to exist in a not so congenial neighborhood, finds it necessary to modernize its defense forces and defense research establishments to world standards through direct imports and collaborations. On the other hand, India's indigenous Defense Research and Development Organization (DRDO) and industrial sector have advanced technical capability that could make them productive partners for US industry in this market. Though these macro-economic and industrial characteristics are recognized and understood by both countries, there are some important obstacles that remain, between India and the United States, which need to be overcome for better defense cooperation. While some of the political hurdles have their origin in the indirect side swipe of the Cold War rivalry, the most persistent ones owe their existence to long-standing bilateral concerns. Virtually all of these obstacles are ultimately reflected in the process of the transfer, control or management of defense technology.

Most analysts agree with the fact that we are entering an era in which there are significant changes forthcoming. These changes are related to the momentum of the Indian sub-continent played against the flash points in that region and global political turmoil. The overwhelming importance of the change in world affairs predicts a significant influence on technology development, technology transfer and the management of this transfer in most developing countries. Though the technology community in the United States has played the role of a leader in most areas in the last five decades, it will now have to deal with new situations whether it be in its internal arrangements or in its international relations. Today, it is a question of being first among equals. Developing countries do not want to be left standing still. These new developments will involve new thinking and new partnerships in technology updates whether it be with the US government, private sector or the academia.

'Men are only as good as their technical development allows them to be' according to George Orwell. One key aspect is that technology is considered today as the most important factor in economic growth underlining Werner von Braun's statement that "A third-rate technological nation is a third-rate power, politically, economically and socially." The truth about commercial and military technologies is well known - that is the protection of a country's "crown jewels." Technological interaction between the West and the East seem to follow a predefined dictum. To avoid losing its cutting edge, a developed country must never part with current technologies. It must sell an older version and, if it is unavoidable, it must share selectively even with allies. One must use the machinery of export administration and control regimes to regulate international technology flows. If this is the norm followed by industrialized countries, even legitimate technology transfer to developing nations becomes a monumental task.

In their search for state of the art technology, developing countries and some developed countries have often been accused of 'economic espionage'. There has been quite a bit written about this modern day 'economic espionage'. But what of the practices of developed countries ? Can there be a bigger impact than the economic espionage than that carried out by Francis Cabot Lowell for the US (Fialka, 1997) ? Is it not well known that Eastern traditional know how has been systematically exploited by foreign powers ? The hundreds of patents mushrooming around the world based on eastern traditional practices is proof enough. This is especially true for pharmaceutical patents. Does the Indian neem ring a bell ? Perhaps the historical naiveté of developing countries in protecting their own intellectual property is to blame. Perhaps the tropical climate has tempered their trust in the fairness of their fellowmen. But these issues are left to more knowledgeable duelists and are not considered to be within the scope of this book that intends to talk of only legitimate defense technology interactions .

It is true that technology cannot be kept static, suppressed or locked away. This fact is felt in the daily lives of technologists, economists, and politicians in developed and developing countries. The cost of being in the technology game is increasing steadily due to rising cost of research, development, testing, and evaluation, manufacturing and marketing in a competitive world. If one adds to this basket the insistence on technology transfer by developing countries, altered security "realities," shrinking defense budgets, the push and pull on the economic stage, and the steady growth of industrial infrastructure around the world, then one is left with the inevitable result of a loosening of technology protectionism (Santhanam, 1992).

Technology transfer has long been a problem in defense cooperation. According to Pentagon officials, the effort to re-examine US policy on technology transfer is highly sensitive and unlikely to result in any rapid change (Hitchens, 1996). Today there is a tendency to liberalize trade with former adversaries and restrict trade with current allies. This is a very confusing state of affairs since, as Nau says, one can lose a sense of the overall strategic and political context in which trade and technology policies acquire their meaning (Nau, 1992). To sort out this confusion it would be helpful from an academic view point at least, to review the circumstances under which strategic export controls emerged in the United States and other Western countries after the Second World War.

Countries will always take sides in any situation depending on perceived threats and opportunities. They will define and redefine their positions based on their previous experiences and what they would like to project of themselves in the global environment. These definitions that are bounded by the concerns of each nation are day to day practicalities that explain how nations behave in their environments. The Cold War was the result of one such perception that was egged on by the military and economic imbalance after the war. The US and its allies knew that the Soviet conventional military power was greater on a one to one basis and could be met only with an input of tremendous resources. So the crux of the cold war strategy was to answer Soviet military advantages by a threat to escalate future conflicts to greater levels of military engagement, which could culminate with the utilization of nuclear forces. This strategy was followed since this was where the United States had a superior capability based on more cost-effective strategic technologies. The United States and its allies had to therefore maintain the leadership in these technologies and they had to find a means to protect it. This was the rationale that initiated export controls and the Coordinating Committee for

Multilateral Export Controls (COCOM) and had the specific purpose of protecting American and West European leadership in strategic technologies.

COCOM was used to set in motion export controls that could delay the movement of critical technologies to the Soviet Union until they were no longer force multipliers or a threat to the defense strategy of the western powers. These export controls soon applied to other countries including the developing ones aspiring for new technology. However from the multilateral platform, the export control regime did not live up to the expectations both in terms of export controls enforcement or in receiving cooperation from non COCOM members. So, while there was general consensus for a regime, there was no consistent philosophy or orchestrated program to enforce these controls. It is common knowledge that the western countries have disagreed about the actual substance of the controls and also about the procedure for implementing and enforcing the controls.

Over the years, plans of action have changed in the West and other export regimes have sprung up and they have all had their impact on developing countries like India for whom export controls are always a burning issue. To understand such a broad canvas, one has to analyze the present international scenario which involves the strategies of the United States on export control, transfer of technology, sheltered high technology, her arms sales and the reason why third world countries demand these commodities. It is also essential to study India's approaches to these issues with the ultimate goal of facilitating technology transfer and its management with relation to Indian defense laboratories and industry so that one can come to grips with a situation that seems to be getting, if not already, out of hand. This realization calls for an overall picture of certain issues like the Missile Technology Control Regime (MTCR), Non-Proliferation Treaty (NPT), the Comprehensive Test Ban Treaty (CTBT), the resultant controls on exports and other embargoes that have impeded technology transfer at a time when there is a substantial reduction in demand for armaments in the world market and vendors are desperate for markets. India is one such market and needs to acquire some of these commodities. Working equations will have to be found as to how US-India bilateral technology transfer in defense technology can be improved and what steps have to be taken in an attempt to find new approaches to technology transfer and its management in an Indian defense environment that is striving to be self-reliant. This joint exercise into a mutually beneficial future must be based on Indian and American attempts in finding common ground.

In the ensuing study of the issues involved with regard to export controls and the transfer of technology to India, certain terms must be understood in-order to avoid semantic confusion. The Indian Ministry of Defense (MOD) refers to the Indian Army, the Air Force, the Navy, the Defence Research and Development Organization (DRDO), Defense Production (DP) and the Ordnance Factory Board (OFB). The Defense Services refers exclusively to the Army, Air Force and the Navy. The DRDO comprises of fifty three laboratories and establishments that work on all India's defense technological research and development. The term Indian R&D includes general R&D activities in other scientific organizations, the private sector in India and also includes DRDO.

R. J. Augustus

2

The New International System

The world in which we live has been turned upside down in the past few years. The turmoil that has characterized world affairs since the collapse of the Soviet Union poses significant challenges and opportunities for India and the United States. The break-up of the Soviet Union may be the introduction to trying times ahead since with the collapse of the bipolar world, we seem to be once again facing certain feelings of 'nationalism' that was so destructive during the earlier centuries. This collapse of the bipolar world has not only left many in the United States wondering about the purpose and limits of American power, but it has also left India wondering about its defense future since it had such close relations with the Soviet Union. General Shalikashvili explains that for the last five decades, the world had become accustomed to a great strategic consistency and says that although the cause of the consistency was deplorable, the fact of the consistency was reassuring and oddly comfortable to all concerned (Shalikashvili, 1995). This era of strategic containment was thoroughly understood and uniformly accepted. This strategy used by the United States succeeded, but when the Soviet Union broke down, the US seemed unprepared to cope with the conditions and consequences they had struggled for so long to create.

Today this collapse of the bipolar world has brought forth new nations and new borders and has created a political tornado that has engulfed every nation. For the United States, many of her allies who were once willing to sacrifice some national interests to help preserve her leadership in the struggle against communism appear less willing to do so now since the threat seems to have abated (Widnall, 1995-A). This leads us to the question that, as the only remaining superpower, to what extent is the world's burdens that of the United States ? To what extent should she get into regional feuds and turmoils where she is probably not welcome ?

Even as nations try to come to grips with their respective political challenges, they are being torn apart by two other revolutions that can be termed as economic and military-technical. As Alvin and Heidi Toffler have described, the world is entering a "third wave" of transformation. This third wave is the information revolution, and it has started to change societies just as dramatically as the agrarian and industrial revolutions before it (Czerwinski, 1996). Czerwinski adds that any one revolution would have been good enough to keep a state totally engaged, but the simultaneous occurrence of all three raises the possibility that we may be entering a long period of instability in world affairs.

The economic and social transformations that the information revolution is bringing with it will forever change our known world. It is widely felt that in the past, while nations had some control over natural resources which sustained the growth of the industrial revolution, today they do not have adequate control over information and the associated technology which sustains today's third revolution. The most profound implication of this may be the growing economic interdependence this information revolution is bringing to the international community.

As individuals and as nations, one needs to be increasingly aware of reactions and counter-reactions around the world.

Seemingly minor events in the Western world or the Eastern world can have dramatic effects on the economy and the lives of each and every one . Every nation must be concerned about world events, because whether they choose to admit it or not, world events will affect them. Developing countries specifically must take advantage of these opportunities and shape the world of today and tomorrow by staying engaged in world affairs. They have to take cognizance of the fact that they must cope with a wide range of potential conflicts all around them including threats that could come from many different directions and in many different forms at any time. They must face greater challenges than ever before to provide for their national security and the potential for world events to directly influence them will grow as new and sophisticated weaponry finds its way into their neighborhood. So countries are finding that new threats have increased and these threats define the need for fast, flexible, mobile forces with the most advanced weapon systems. The onus is on the respective governments then in the face of these threats to ensure that their fighting forces are prepared for what is demanded of them. This would then call for equipping the defense forces with the most technologically advanced products that are affordable and accessible to the governments. Hence one can come to the conclusion that technology is the foundation of any defense strategy in the post-Cold War era as it is a completely different ball game. Paul Kaminski explains that the United States, as the only remaining superpower, feels that it faces scenarios that involve regional conflicts with diminished potential for direct military threats to its territory. However, there is increased likelihood of its forces being committed to limited military actions in which allies are important partners (Kaminski,1995-A).

R. J. Augustus

Though the future looks bleak, it is not all that dark. There is growing international cooperation on many fronts. With the progress in reforms in the Soviet Union and Eastern Europe, starting with 'Perestroika', the East-West relationship has changed from confrontation to cooperation (SECC, 1993). However, to the pessimist, the world remains a very complicated and potentially dangerous place. The key feature of the political, economic and military changes is that they add uncertainty both for decision makers in the United States, in India and those in other countries. In addition to uncertainty, however, these ongoing changes provide opportunity. Sheila Widnall feels that what nations make of these opportunities will depend upon how proactive each one is and how engaged they remain in international affairs (Widnall, 1995-B). To really understand why nations do what they do or what they have to do or ought to do, it is necessary to understand this "new world order" in which we find ourselves embroiled.

In today's world, developing countries do not have the option of isolating themselves. World events will affect them and so they must stay engaged internationally if they hope to pro-actively take part in and shape world events . The threat of worldwide nuclear conflict is greatly reduced with the end of the cold war, but the threat of regional conflicts has grown. These regional conflicts are proving to be greatly expensive in terms of human life and destruction of the economy of the countries involved. One can see in today's conflicts that the craving for economic advantages or misguided nationalism in congruence with modern weapons, is causing untold damage and suffering to all game players. To compound the threat, nations are seeking to sell and acquire all sorts of offensive weapons to serve their own agendas. Because of this changed security environment, nations are starting to rethink their regional strategies especially with the nuclear ambiance. In the United States there is a rethinking both in the context of keeping the threat of a global nuclear confrontation at a minimum and in preventing regional conflicts (Perry, 1995-A), for, no sane country or person would risk being part of a nuclear attack - a mortifying scene so vividly described by Blackwell and Carnesale (Blackwell & Carnesale, 1993). The avoidance of such confrontations then calls for interaction on a different plane between countries that do not see eye to eye. One administration called it shaping a new world order, even as the Clinton administration refers to it as "engagement." The fact is that "engagement" does have a very real and a very vital purpose especially when one looks at it from the Indo-US angle, as in this book, where there is an Indian and US manifest for creating new possibilities and new opportunities.

Changes In The Military Threat

The discontinuity that has occurred in world affairs has significant consequences for American and Indian defense policy. Since an effective defense strategy depends to a vast extent on the application of defense technology to meet the needs of the armed forces, these changes in the world imply changes for defense technology and therefore they directly affect the overall functioning of the armed forces. So the effect of the changes will be to enhance rather than to diminish the importance of technology in the framework of Indian national security. What must also be considered is that, based on the present world situation, the US Secretary of Defense has come to the conclusion that the military threat to the United States has lessened (CCSTG, 1993). The US defense budget has decreased over the last four years, with the expectation of further reductions and reallocations in keeping with the decreased threat. This trend will continue for the foreseeable future if the Soviet Union maintains its current foreign policy and also if the East European nations sustain their move toward noncommunist governments independent of the Soviet Union. If this continues, then there will be a decrease in the defense requirements of the United States and hence a reduction in US defense expenditure.

As for India, her dependency on the Soviet Union for close to four decades is now jarred. India, based on her own thinking and US attitude, depended on the Soviet Union for most of her weapons and today, in some circles, it is considered as a handicap that is causing problems in international relations. So India must now think of correcting this by identifying alternate sources of supply for building up her much neglected defense capabilities. There could be a change in focus towards the United States, but India seems to be hesitant. Of course, India and the US do not find their task any easier if one only listens to the cold war warriors on both sides. Whatever be the Western opinion, India's threat perceptions have not gone away. India feels that she alone has the responsibility to provide for her own security.

However, looking further into the future, there is a general feeling that one cannot discount the rise of other major economic and military powers, some of whom may not be friendly towards India or the United States. So, if one pushes for jointness in defense cooperation between India and the US, their future military requirements will be demanding, even though it may be quite different from what was expected from these countries during the cold war or India's problems with her neighbors during the last few decades. When asked to identify the threat to which US military security should be directed, President Bush answered, "Unpredictability, uncertainty, and instability." Technology, therefore, from the US point of view, is an important insurance policy against an uncertain strategic future and needs preserving, and broadening of the defense technology base. This direction of thought in the face of a reduction in overall defense spending is an example of how defense policy is influenced by global events. In this regard, India should have and has learnt a lesson from the US experience.

Why Proliferation ?

One of the dogmas of international relations is that countries act in accordance with a perception of their interests and power. A nation's actions and policies clearly reflect its interpretation of the dangers prevalent in its environment and triggers responses that are considered necessary to protect it against these perceived threats. 'Proliferation' and 'rogue nation' are words that are frequently used in the arms transfer business, but the irony is that who they refer to depends on which side of the fence one is on. So what sort of forces or threats could cause a nation to go in for serious weapons ? Why do some nation's feel so threatened that they consider the need for 'proliferation' of weapon systems in the face of global opposition ? These and other questions need to be addressed for a better understanding of the motivating factors. Though any type of proliferation that might perpetuate global harm is discouraged, for the sake of analysis and priority for global protection, the worst case scenario of nuclear proliferation is considered.

Firstly, each country will make the decision to possess nuclear weapons because it is convinced that these weapons will improve its overall security. Martel explains that the history of the Cold War demonstrated that possession of nuclear weapons directly increased the security of the United States, Soviet Union, China and West European states. This was mainly because costs of war were far higher than any envisaged gains (Martel, 1995).

So one comes to the question whether international security is really enhanced by the possession of these weapons. The tenacity with which the super powers had gone about the development and possession of these weapons earlier and the reluctance to fully give them up today and the penchant developing countries seem to have toward them, seem to indicate that there is a consensus to this extent. One must also notice that those countries with an 'established practice' in nuclear activities now term as 'proliferators' those countries who aspire to be on par with them. A strange turn of events, but which cannot in any way excuse the gambling with nuclear weapons and human safety. So whether global proliferation of nuclear weapons portrays an agreement among many countries that nuclear weapons contribute not just to national but also to international security in today's scenario, is a matter of debate.

When a nation decides to develop and possess nuclear weapons one hopes that it is done so only after that government has engaged in a careful and prudent consideration of relevant political, military, and economic factors. They should clearly understand that just as nuclear weapons can enhance their security, they also can heighten tensions in their region and increase the chance of a nuclear war. The stand taken by the present Indian and Pakistan governments to keep their nuclear options open is a case in point. However in all fairness it must be acknowledged that each country is the best judge of its own security interests and the power necessary to protect those interests. These nations are uniquely qualified to evaluate their interests, the probability of other nations interfering with them and the ability to defend their interests. These nations also take the stand, perhaps rightly, that foreigners to that region, do not have the capability of evaluating all the concerns and are therefore not qualified to pass judgment.

Admiral Turner in his book ' Caging the nuclear Genie' puts forth his ideas to show how nuclear proliferation is posing a challenge to America for maintaining global security and mentions that countries like Israel, India and Pakistan have all shown considerable caution in controlling their nuclear capabilities. He also mentions that there could be the possibility of nuclear weapons being used and in that case the United States would have to step in to bring the nuclear conflict under control. He frankly acknowledges that such a policy is not politically feasible since if the US espouses the use of nuclear weapons to defend its European allies against conventional attacks, then they could hardly deny any other country the same right. So the only alternative is to renounce first use (Turner, 1997). Because states carefully weigh their power in comparison with that of other states, they are attuned to imbalances that weaken their ability to serve those interests. Nuclear ownership by India or any other country therefore, constitutes a tightrope exercise in balancing their strategic interests with the power they can summon to defend those interests.

R. J. Augustus

Post-Cold War Reality

The advent of the post-Cold War era has not had such an impact in Asia as it has had in Europe. Yet, a sea change is clearly visible and nations that were at war or in the midst of tense relationships are forging new links. The political, economic, and security environment in the region is being transformed by a general process of détente even though some of the changes had commenced even before the Cold War drew to a close. But the end of the global conflict and competition between the groups of countries led by the United States and the Soviet Union has brought about a realignment of forces unimaginable a few years ago. Today it has forced virtually every nation to reassess its domestic and external strategies and draw up a new agenda based on its own priorities.

Dramatic improvements in India's relations with the United States and China are not only major bilateral achievements in themselves but carry wide implications for other relationships. It is hoped that one area where their impact is likely to have a positive outcome is in Indo-Pakistan relations. In the subcontinent, India's close ties with the Soviet Union kept the United States aloof. S. Dutta explains that this divide was used by Pakistan to build up its defense capabilities against India. The strategic alliance between China and Pakistan and China's attitude toward India in the 1960s was instrumental in Pakistan's challenges to India and the subsequent wars. It is felt that 'Pakistan could not have pursued its nuclear weapons and missile programs without significant Chinese support and less important, but nonetheless significant, tacit US understanding' (Dutta, 1992). Again, the recent testing of Pakistan's liquid fueled 'Ghauri' Missile, whose design is supposedly based on North Korean technology (Weiner, 1998) and the proposed testing of the 2000 kilometer range 'Ghaznavi' is sure to muddy the sub continental waters. However to optimists like this author, it is hoped that the Asian environment will gradually change for the better even though there is some saber rattling to be heard now and then. The United States is a key factor in this process of change and is actively involved in reworking many of the past politico-military ties and forging new ones and is a key global participant in the search for a solution to the regional conflicts. But the changed context has not eased the pressure on those involved and the concerns felt. New security perspectives are taking shape and are bringing with it a new thinking in terms of defense self-reliance and the need to seek protection with the latest in weaponry and technology.

Third World's Demands For Weaponry

Throughout history, every major power that has dominated the world stage has had a world-class military structure to sustain its power. This is true today of the United States and its defense services. Over the years the US has always been a pace setter in defense technology. In the 1940s the United States was the world's preeminent technological leader and primary source of such advanced defense products. By 1990 the United States was competing in high technology products against producers in Japan, the European Community, and the newly industrializing economies. The United States declared its first high-technology trade deficit in 1986 and has since struggled to maintain a small surplus (Long, 1992).

However, Fogleman clarifies that to maintain a technological lead over the others, the US has already begun working on the next family of high tech defense equipment and one will soon begin to see the evidence of the breakthroughs in the areas of digitized battlefield, information warfare, the new generation warrior and so on. With all this talk in the air, third world countries are beginning to look for means and ways to modernize and replenish their inventories (Fogleman, 1996). They are now looking into the future through US eyes and capitalizing on US experience instead of reinventing the wheel. This is especially true in the case of Information Technology where every establishment in the US is latching on to the ongoing revolution in information technologies and third world countries are following suit. The flexibility and speed of today's technology being used in the area of defense make it uniquely capable of utilizing real-time information to dominate someone else's battle space. Keeping abreast with this technology explosion is particularly attractive to developing countries.

In this context, the Monterey Institute of International Studies has identified India as an emergent missile power, which has devoted significant financial resources to build rocket expertise and is now ready to enter the select club of states having both a robust civilian space industry and viable "missile-based" defenses. With its increasing capabilities to design, build and test missiles and nuclear weapons, India is emerging as a 21st century world power, whose strategic policies must be given due weightage, by the more mature countries. It is felt that India could also emerge as a potential supplier of sophisticated missile technologies. It is feared that if countries like India are not careful with regard to their export controls, it could adversely affect the efforts of the developed countries. However, in spite of projected capabilities in defense technology and defense industry, India, like other developing countries realizes the necessity for modernization of her defense inventory and technology and at the moment at least, looks to the more developed countries for help.

New Technology - The Need to Modernize

If one steps back to look at the broader canvas, one can recognize that each and every country is going through an enormous debate about its institutions. Like the US which has found that they are in a situation where they need to change, so too are the perceptions of third world countries. J. White (White, 1995) explains that with the change in the world scenario, once known threats have now changed, the world economics have changed and the priorities of the man in the street has changed. Those institutions that had provided each nation with good capability over the last many years have now outlived their usefulness. Recognizing the need for change, governments all over the world are realigning their goals and objectives and are trying in several ways to revamp their defense establishments, their defense forces and their defense inventories. India, also caught up in this cauldron of change, is finding that these changes are not that easy. The critical issue has been in identifying where India has to go and to how to make that change so that Indian defense forces and Indian defense R&D can reach their goals of enhancing their capability and operational efficiency in spite of several constraints within the country and outside its boundaries.

There are several motivations that regional countries have for acquiring modern military capabilities. The largest motivation is a response to legitimate concerns about external threats to security. In some cases there may also be concerns about maintaining internal security for which high performance military capabilities are probably not required to meet that threat, but it can be a great booster to personnel morale. Modernization of the defense forces being the most important factor, the defense services need to be equipped with the state of the art technology so that when they have to fight, they have a good chance of winning. India learnt a very bitter lesson during the Chinese aggression. So in equipping these forces, minute attention has to be paid to what they need for the planned mission. This fact has not been lost even in a high technology US defense environment for, while addressing this all important detail, the US war fighting commanders in chief during one of their meetings spread a very significant message among the bureaucrats saying "Don't forget the small procurements. Don't forget the things that make a difference in combat capability" (White, 1995).

India like any other developing country has an urgent need to modernize her forces and the way modern warfare and defense is moving, all of this new capability has to be joint capability, because joint operations will be the style in the future in almost every case whether it is a national or an international endeavor. If an international endeavor is called for, then it is absolutely necessary that the technological capabilities of all participating forces be of the same standard. So if India and the US face a common threat, it is imperative that compatibility of battlefield technology is ensured. Jointness takes a lot of training, and it takes a lot of cooperation and trust among countries and their services. While this is extremely important, it is not always easy to put into practice since modernization means a truly joint vision and not merely new weapons. This then would call for outlining a new doctrine for joint planning, training and operations. However, one very real fear of the US is that at no time would they like to find themselves facing an enemy who has the same technology and the same training. If this is a fear that colors Indo-US relations at any time, then an analysis of the Indo-US relationship has to be done to allay any such fears that Americans and Indians may have of each other. Any Indo-US cooperation in defense has therefore to be a win-win situation based on understanding of each one's view of the matrix of situations.

Arms Race On The Subcontinent

The words of John F. Kennedy "we dare not tempt them with weakness; for only when our arms are sufficient beyond doubt can we be certain beyond doubt that they will never be employed" seems to have had a telling effect on the people of the subcontinent. The saber rattling that India and Pakistan have been doing and the vows taken by each country to outdo the other in the area of defense gives one the feeling that a Cold War has started all over, this time in South Asia (Cooper, 1996). With the two countries engaged in a long-standing rivalry over the territory of Kashmir, the Central Intelligence Agency (CIA) in a recent study ranks both India and Pakistan among nations considered to be most at risk of serious instability and warns that the world's greatest potential for nuclear conflict lies in this region. This CIA prediction has been consistent over the last few years. In 1974 after India conducted its first underground nuclear test, the then Prime Minister Zulfiqar Ali Bhutto declared that Pakistan would 'go for nuclear status even if we have to eat grass.' Since 1987, Pakistan has said it possesses the know-how and material to make nuclear weapons, but it has yet to demonstrate its capability by testing one. Pakistan feels that this is a good deterrent and Cooper quotes Rafeeq Afghan, editor of Takbeer, a widely circulated news magazine published in Karachi as saying "That's why we have adopted nuclear capability, for a deterrent; we thought this deterrent was very effective in the Cold War between the Soviet Union and the United States because they never directly fought a war (Cooper, 1996). In January 1996, India carried out another test of its new Prithvi missile with a range of 155 miles, enough to reach Pakistani cities such as Lahore, Islamabad and Rawalpindi. Pakistan vowed not to be outdone and Foreign Minister Sardar Asti Ahmed told Pakistan's parliament that "If India wants to prove its manhood by conducting a test, then we have the capability to prove our manhood." Perhaps Indian strategic planning is at fault. Once having conducted the 1974 nuclear test, India felt the brunt of international pressure which tried to curb the development. Pakistan made use of India's test to seek an excuse to publicly announce their nuclear program. So having lost out on international cooperation, why could not India go ahead with other tests ? Why did India have to deny her defense forces the option of nuclear weapon integration as part of their operational planning ? India has only achieved the distinction of being termed a nuclear capable state. If the technology was pursued with a series of tests, India would have had an abundance of data today. Should India now pay for its consideration for world opinion ? This is a question that is coming up in almost every forum debating Indian defense strategy where today it appears that the good get punished and the bad get rewarded.

The sub continental arms race till date has caused complications in US relations with India and Pakistan alike while creating better business for arms exporters. In addition, it has created an extra irritant in relations with China, a country the CIA contends has supplied nuclear-related material to Pakistan. Based on the Pressler Amendment, the US Congress has tried to discourage Pakistan from developing nuclear weapons by cutting off US aid in 1990, including conventional weapons suppression. Pakistanis argue that the ban actually boomeranged and compelled Pakistan to pursue nuclear capability as its only remaining option for self-defense against India. The US congress in November 1995 approved the one time waiver of the Pressler Amendment so that Pakistan could receive reconnaissance aircraft spare parts and other materiel for which it had paid $368 million. But now that exemption has been bound up in the debate over China's nuclear-material deliveries.

Today the rockets fired from India are built by Indians and can easily hoist satellites into polar orbit. With its most modern Prithvi missile, it is felt that India will be able to place a nuclear warhead on a target nearly 200 miles away (Zimmermann, 1996). Pakistan appears determined to "match whatever India does,' even though Pakistan already devotes 35 percent of its budget to defense and 25 percent to interest payments on money borrowed largely for military purposes. The recent 6 April 1998 announcement of the testing of the 'Ghauri', a missile capable of placing a nuclear warhead more than 900 miles into Indian territory proves that Pakistan's determination. With the 'Ghauri' in the news and the alleged Chinese help, there has been a prompt response about India getting Russian help to build its 'Sagarika' missile ! (Myers,1998).

So now, no one can or needs to raise the question of sanctions against China or Pakistan.

For more than two decades, a slow arms race seems to be creeping across the Asian subcontinent. Indian and Pakistani nuclear capabilities have been hinted at by the respective governments and the rest of the word is allowed to draw their own conclusions from the ambiguity. Now, as new missiles and nuclear technology work their way into the subcontinent with China hovering in the background, the alleged nuclear programs of both India and Pakistan are painting the region as one of the most dangerous areas in the world. While the United States and the Soviet Union were a great distance apart, India and Pakistan have a common border which speaks for the proximity. Not a very conducive range to use nuclear weapons for it would anyway be a lose-lose situation.

US analysts believe that China cut the ribbon for Pakistan's nuclear weapons program by providing a bomb design and some weapons-grade uranium in the 1980s. They are also helping Pakistan in the construction of a secret plutonium production reactor at Khushab. According to the Risk Report, published by the Wisconsin Project on Nuclear Arms Control (Risk Report,1995), the reactor will eventually be able to produce key ingredients that Pakistan needs to build more-sophisticated, powerful bombs. So Pakistan, a narrow country that is vulnerable even to short-range missiles, is struggling to keep nuclear pace with the rest of the world. China and Pakistan have twice been hit with US economic sanctions for the transfer of Chinese M- II missile technology to Pakistan. US intelligence analysts are convinced that entire M-11 missile systems, which are comparable to the latest version of the Indian Prithvi, have been shipped to

Pakistan and are being hidden in crates at Pakistan's Sargodha Air Base. The Washington Post, citing a draft government report, says that US intelligence agencies had unanimously concluded with ``high confidence" that Pakistan has obtained M-11 ballistic missiles made by China (Velisarios, 1996) and also that Pakistan has probably finished developing nuclear warheads for these missiles. Though the Clinton administration has irrefutable evidence that the M-11s are in Pakistani possession, no sanctions have been seriously imposed against both Pakistan and China. Finally it is also doubtful if the International Court of Justice ruling on the illegality of nuclear weapons (NYT, 9 July, 1996) will ever be heeded by India and Pakistan or for that matter by any other country with serious threat perceptions.

R. J. Augustus

3
 Anatomy of Indo-US Relations

India and the United States are at a momentous time in history and can change the whole way they look at each other in all the different areas of interaction and cooperation. Figure 1 represents some of the broad areas of interest in the bilateral relationship. Both countries are at a historical stage where a special and close relationship can evolve between them to mutual advantage, if only they seize the opportunities that are now open and if they manage skillfully to contain the issues that needlessly could divide them (Bhagwati, 1994). A proper analysis of their relationship would throw things into a better perspective knowing that every situation whether political or scientific has at least two sides to the picture. This cannot be more true of the Indo-US relationship. " A common faith" is the way Jawaharlal Nehru described the Indo-US bond and Dean says that it is a truism that India and the United States are the two largest democracies and yet, behind the cliché, there is deeper significance (Dean, 1987). Though the United States presents a picture of acting as the global policeman, this is not necessarily true or if it is, it is bound to change. However, the US would definitely intervene anywhere when its interests are perceived to be at serious risk. The US has an opportunity to take advantage of its Cold War victory by encouraging world peace and stability and it is bound to do so while being selective about foreign involvement since economic interests would be a major determinant.

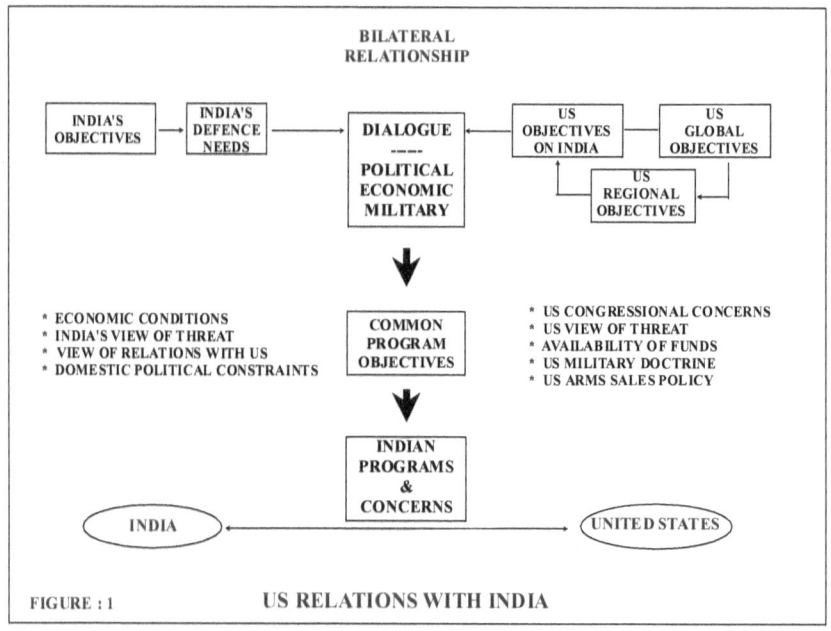

FIGURE : 1   US RELATIONS WITH INDIA

India and the United States have a lot in common with each other. In each country there was a determined struggle for independence led by Mahatma Gandhi in India and by George Washington in the United States. The democracy in each country depends on institutions which are either British in origin or which developed from the British as the two greatest democracies of the modern world emerged from the experience of British rule, as well as from gaining their independence from Britain (Rees-Mogg, 1996). So they have a natural affinity, having shared the same colonial master and which gained them English as the unifying language. India and the US have been democracies from their independence, India for about half a century, the United States for two and a half centuries. Though democracy has spread around the world today, one must remember that India is studied as a great example to the lack of democracy earlier in the Third World. Talking of ideals, Mahatma Gandhi borrowed from Thoreau the notion of peaceful civil disobedience and Martin Luther King borrowed this ideal back for his civil rights struggle. All of this puts the two countries on a similar platform.

The Indian community in the United States today is over a million strong and this immigration is deemed by excellence of the highest kind in virtually every walk of life in the US. In today's world, India's intellectual contribution is highly valued and plays a very significant role in attracting US attention. One must also take into consideration India's population, relative size and surging economy which must make the US extra sensitive to India's concerns in her own region. Americans till date tend to equate India with Pakistan rather than with China. Indians feel that their importance should be far greater relative to that of Pakistan and Bangladesh. So to the Indians, being equated to these countries, by the Americans, is considered immature if not mean.

There is a general feeling in American circles that Indians think of their influence in the region as greater than just the subcontinent. Indians on the other hand ardently believe that the American government deliberately defines their region of influence too narrowly. During the Third Indo-US Strategic Symposium it was felt that 'India sees her area of maritime concern and desired influence stretching from the Persian Gulf at least up to the Straits of Malacca and on land including China' (TIUSS,1992). The same opinion is voiced by Khan (Khan,1995). India has regional concerns, not ambitions. So treating India only as a South Asian power relative to Pakistan, may not be entirely accurate or acceptable. One must also keep in mind the long time Indo-Soviet friendship as one of the major irritants in any Indo-US relations. Historically, Soviet-Indian relations have developed steadily over the years with the 'Delhi Declaration' reflecting the unique nature of Soviet-Indian relations based on principles of sovereignty, equality, non-interference in others international affairs and cooperation (Gorbachev,1987). So during the Cold War, this contrast in relations with the Soviet Union made the Americans feel let down or betrayed. There could be a danger of new irritations that might replace the old. American tolerance of Pakistan and China's activities in the South Asian arena and the technology support being provided by the present US administration to China might soon erode the reasons for keeping the two nations linked together. From the American side there is a perception of Indian nuclear proliferation and violation of human rights that could be the spanner in the works. The US must make the effort of understanding the Indian situation and the reports that originate from the subcontinent. After all India too is and has always been a democracy with a free press, a strong judiciary and with a powerful, independent and assertive electorate as proven even by the recent 1998 elections.

When one considers the influence of technological development on the relationship, it is obvious that technology will be one of the main hurdles, for both nations, against the uncertainties of the future. American defense technology has been directed for almost half a century against a more complex and uncertain enemy, with American dominance in virtually all fields of technology, especially defense technology during the post war period. Today the direction is different as this dominance is giving way to a position of first among equals. As the world grows smaller, and the distinction between "defense" and "commercial" sectors of the economy merge, India and the US find themselves more involved with each other on many different planes. Yet both countries are yet to come up with a blue print to unite them in a common strategy. How does one try to kick start a strategic convergence since both countries are not going to dramatically alter their policies ? Where should one attempt to look for common ground ?

India's defense establishment is also limited by economic realities and defense spending has declined steadily over the past few years. Yet, Indians hear constantly about India's growing power projection capabilities, even though no new systems have been introduced or are likely to be, given continuing fiscal realities. There have been accusations that India has amassed weapons and armament from the Soviet Union. This is far from the truth and if one looks at the actual sales to the Indian armed forces and the amount of equipment that has become obsolete, one will arrive at a clear picture of what India really has for her size and situation.

Indians and Americans agree that the non-proliferation issue presents the greatest impediment to improved Indian-American relations. While Americans constantly urge India to join the group of nations seeking to curb proliferation, India seeks an American recognition that only a nuclear restraint regime could be realistically applied in South Asia. Although most Americans accept the rational for nuclear restraint, rather than a nuclear weapons-free zone, the latter remains the American policy goal for the region. So this issue of nuclear non-proliferation is a very sensitive subject in the subcontinent and again needs to be understood by the US. India has resisted the Non-Proliferation

Treaty because it was considered as completely asymmetric in responsibilities and obligations between the small and the large powers. India also considers it to be biased between the members and the non-members of the Nuclear Club. Over the years, India has objected to this treaty simply because Nuclear Club members are exempted from certain obligations while the non-members have it imposed on them. From where India comes, she feels that it is prudent to have a multilateral curb on nuclear weapons for all countries. But, in her own interest, India is likely to explore regional options as well, to prevent nuclear proliferation. During this process the US role will have to be by invitation and not by undiplomatic meddling. If this is not adhered to, as has happened several times in the first term of the Clinton administration, outright negative comments will produce a hostility in the Indian public against America and opportunities that could draw the two nations closer will be lost.

Let us consider the tremendous impact that India's economic reforms have on her present stature. It is believed that Indian democracy together with economic reforms will give India a future that will outshine the advantages that similar reforms have given to the Chinese. Economists have predicted that the economic growth in Europe and Japan will be much slower over the next two decades than markets in the rest of the world and also that the world economic growth is shifting to places such as Brazil, Argentina and Mexico in the West and to China, Indonesia, South Korea and India in the East. Growth rates in these countries are expected to be phenomenal and for many like India 7.0 %, Indonesia 7.5%, Singapore 8.2%, Thailand 8.7%, Malaysia 9.6%, South Korea 9.0% and China 9.5 %, this demands notice at a global level. India like many of these countries is instituting significant economic reforms, opening up to international markets and the pursuit of high technology. The Indian giant is awakening later than the Chinese giant, but it will step onto much firmer ground. This opens up an unparalleled opportunity for United States businesses to get into the vast and growing Indian market. As with all bureaucracies, results often take time to come in, but they surely will. US businesses are already getting into India on their own and with Indian coaxing. Such a fruitful relationship calls for Indo-US interdependence. However one glitch as perceived from the Indian corner is the Pakistan-China-US triangle that presents itself repeatedly in all Indo-US discussions. This tends to jaundice Indian relations with the US as it calls for US rethinking on its relations with Pakistan.

The different Indian and Pakistani approaches to foreign relations have been reflected in their relations with the superpowers. Pakistan, concerned primarily with the perceived Indian threat and suffering from an extreme disadvantage in size and capabilities, has traditionally turned to the West, and particularly to the United States, for assistance. India, on the other hand, has wanted to be considered a great power in her own right and has been reluctant to accept the bipolar nature of the international system. India decided that she had a friend in the Soviets who offered her equipment and technology with not too many encumbrances. So she signed a treaty of friendship and cooperation in 1971 with arms imports primarily from the Soviets. Pakistan on the other hand has received the bulk of its arms imports from the United States since 1981. India also has an extensive defense industry of its own, although it is still dependent on imported Western technology for most high technology systems. Kemp explains how India never saw its relationship with the Soviet Union as that of client and patron and resented any attempt to compare this with Pakistan's relationship with the US (Kemp, 1992). On the other hand, of course, is the fact that the US does have its superpower status, and her own different set of problems. One needs to put these conflicting views into some perspective, if the two countries are to make steady progress. India will gradually grow in military and economic power no matter what the US does (Harrison & Kemp 1993).

In the final analysis Perkovich states that Indians, Pakistanis and Chinese will decide the future of nuclear weapons programs in South Asia and not the Americans or any other western country. Fundamentally these crucial decisions will emanate from far-sighted and deliberate evaluations of national, regional and global interests. Americans and other outsiders can only facilitate and coax the regional players in this process with their own analyses of the interplay between regional and global interests (Perkovich, 1995). Where appropriate, the international community should go further and affect the local national and regional calculations of interest by providing incentives for considering broader global interests. Among these interests are the avoidance of costly conflict, the non-diversion of resources which could otherwise enhance economic development, and the buttressing of democracy and economic reform. Critics feel that if India, Pakistan and China satisfy the global interests, their national interests will automatically be satisfied. So as far as these three regional players are concerned, there is no question about who should decide on what and when. But in real life, they feel that they are being questioned by those who have no right to do so. One must realize that Western and Eastern politics, culture and psychology are vastly different and make things very difficult even for those who would like to sincerely level the playing field. To promote an interdependence among nations one must learn to recognize these impediments and make allowances for them at an early stage in the endeavors.

Interdependence Among Nations

The term 'Interdependence' according to Rosecrance, has so many and varied meanings that it is no longer fully clear what investigators intend to signify when they use the term (Rosecrance, 1982). In a very loose and general sense, one can say that interdependence is a state of affairs where what one nation does has a direct impact upon other nations. In this book, interdependence is used in a creative and positive sense as a powerful force to create better harmony among nations. When nations decide to interact with one another, they automatically seem to define their interdependence as being active and passive. Active nations use many types of force to undermine, maintain or otherwise influence strongly another nation's system of governance while passive nations do not initiate such behavior. Maghroori elaborates on the different models that have been used to portray this interdependence taking into account the behavioral patterns of nations and the extent to which a nation's sovereignty is compromised by outside forces (Maghroori, 1992). In the long run, there have been those who advocate globalism and those who advocate realism. The learned difference of this perception of cooperative security and national security is in itself a sensitive issue because the world scenario is not as clearly defined as we imagine it or want it to be.

R. J. Augustus

Looking at the economic changes happening in nations, especially the Asian trend, one realizes that economic interdependence among nations has increased sharply in the past half century leading to dilemmas for national politics, cross-border spill-overs and diminished national autonomy. However it has also been suggested that those who believed that world interdependence was accelerating have to admit that nation-states have gone to considerable lengths to reshape the economic, technological, and ecological forces acting upon them, seeking to reduce interdependence (Aaron, 1995). So with this state of limbo, it is no longer clear what kind of international economics will be stable or where it will move.

In talking about the interdependence among nations, one might extrapolate and say that the internationalization of trade and business has removed national barriers to the transfer of technology. An example of this is seen in how more and more multinational firms are exploiting their technology globally and to a lesser though increasing degree, are gaining access to new technology through worldwide diffusion of R&D and collaboration. The term techno-globalism emerged in the 1980s to refer to this activity (Ostry, 1995). But the great concern for the United States is that this high-tech trade is moving out of its realm of control. Other industrially developed and also developing countries like India are beginning to play a bigger role in this high-tech industry that has become trans-national.

With these points in view, it would be prudent to analyze the fundamental thinking of India and the US since their strategies and perspectives are molded after their search for security facing demographic, geographic, historic and political constraints. India and the US have their own strategic plans and their respective stands on global issues can only be understood and appreciated by a study of their respective strategies and perspectives. The pre-eminent effort by both countries should be the avoidance of any type of a Cold War and success in this endeavor would definitely call for participation of other nations supportive of this policy.

Role of the US and India

It is widely felt that America's international relations have not been renowned for its long term, balanced strategy. American policy towards Science and Technology has been at cross purposes with the policy of foreign affairs and Nichols states that this absence of proper analysis and policy leads to un-preparedness for major issues for international negotiations (Nichols, 1993) and dependence on its allies have at times painted a dark picture of the US to world. Let us consider what the US is capable of, if it devotes and receives unconditional support both at home and abroad. The US as the most powerful democracy could form a like-minded club of democratic states which could maintain peace. The United States could establish this consensus all by itself because of its unmatched military power, technological leadership and its research and development that fuel the engine that drives the US economy (Gansler, 1995). This effort will ensure that all the resources pumped into the Cold War era were worth the trouble. America and its allies, past, present and future, could prevent a return to the hostile environment that engulfed the world for the last several decades. Again, seen through India's eyes, the close interaction that heralded the multi-national coalition in 'Operation Desert Storm' seems to suggest that the world seems to be shifting towards being multi-polar rather than uni-polar. So the American stress on balance of power in Asia might open opportunities for American-Indian cooperation, provided the US did not attempt to balance or curb legitimate Indian ambitions.

There is no real enemy for the US today and there might be a tendency to isolate herself, but it would be in the interest of the United States to exert its world leadership through cooperation with other powers from different regions. It has been projected in a number of forums that six powers - the US, Japan, the European Community, China, India and Russia will fill the dominant world position previously occupied only by the US and the Soviet Union. This presents India with an opportunity for enhancing her global influence, but at the moment, her single goal remains her own social and economic development. There are a number of good things going for India that might tilt the US balance of support in her favor. While there are several states in the South Asian region that are facing severe social and political challenges, India by the nature of her own diversity is structured to cope with such problems and may be as good or have an edge over China in this regard. Would the Americans prefer to have China at the helm of affairs rather than India ? India is also a leader of the Non-Aligned Movement, which supports the wider Indian and American interests. India and Pakistan dominate the South Asian scene, but from any perspective, Pakistan does not pose any type of a challenge to India. So even if Americans feel that these countries might pose a nuclear problem, India should again from all angles win hands down. So in all fairness to India, she must be America's choice for a long time friend and try to reach an understanding of working together to achieve international harmony.

India and Pakistan are at a stage when their own internal problems are enough to keep them busy for a while. After a period of political instability, India, at least now, after the new government has taken over, seems to be recovering her balance and is all set to jockey herself into a favorable position in the new century. The US should be keen on ensuring this happens. Any country that seeks to do business with another or has an interest in that country will automatically try to influence the machinations within that country so that the path is made a lot smoother for it to work through in order for all or some of its goals to be met. This is but natural and so it should not be too much of a surprise if the American government tries to influence Indian domestic politics to continue with its economic reforms and other paths of progress, for the US has much to gain. Again, the US may wish to steer India's policy towards her neighbors in all spheres from medicine to missiles because that would commensurate with the US interests. But then, depending on India's concerns, this US strategy may not hold water if it is not practical from the Indian perspective. Developing countries are fast learning that the world does not owe them a living and they have to fend for themselves or forever stay silent and India too realizes the 'power of the buck' in keeping Americans interested in the region. India's age old culture, her ancient history and chant of being the largest democracy does not sell very well in today's international market as values have changed and bargaining chips are different. So there is a need for Indians to push for a stronger economy. This will help them to be able to influence American foreign policy towards India. It should help in making Americans understand that India is the lynch pin to South Asian success. However one must also make allowances for the political pressures within each country which dictate the focus of their foreign policy. Should India then bend to some of US and International pressures in order to be accommodated for some of her requirements? While it is believed that India must get closer to the US and stay there, one must also say that if India strongly believes in some things, she should endeavor to hold onto those beliefs. This is a trauma of another sort for India and as mentioned in the opening lines of this book, by nature of her resilience, she must and will get over this trauma too by playing a bigger part in the international order and should get into the global act rather than remain in the stands. After all it is India's ambition to be the winner in her region and the burden of proof as to why she must be termed a winner, lies with India.

Why India Matters

Before we address the specific question of why India matters or should matter to the United States, one should try and understand what Asia and the Pacific rim countries mean to the Americans. Statistically, five of the US states border the Pacific, with Hawaii in the center of the region. Again, more than seven to eight million Americans can trace their roots to this region which is bound to kindle some American interest. Today, Asia is the world's most dynamic economic region with more than fifty percent of US trade. This Asian trade has helped create three million US jobs (Perry, 1995-A). Therefore the US economy has become interdependent with those of Asia and so any thought of withdrawal from the region would be catastrophic to the US economy. This region is now the hub of economic growth and is ready to take over the role of the world's economic engine in the next century. But the Asian progress will depend on how the individual participants work together to create a stable platform for launching their economic ambitions.

General Shalikashvili explains that just as the United States is looking for ways to keep the European nations from forming splinter groups at cross purposes with each other, so too must the US work toward the integration of Asian security. He adds that in this regard, the US must try to create an order 'built on trust, cooperation, mutual interests and interdependencies rather than on ancient competitions and old scars, for in these regions lie the powerful economies that are the engines of future global prosperity and also much of the material and human resources upon which these engines depend' (Shalikashvili, 1995). One should however realize that Asia's historical distrusts remain ingrained and is quite obvious whether you look at the South or South East. Notwithstanding this, the writing on the wall indicates that US interests in the area will dictate her increasing interaction with Asia. In this context, specifically for India, US Senator Charlie Rose says that India has become strategically very important to the United States because of the fact that now China has nuclear weapons (Kranti, 1996).

The US and India are drawn together by virtue of their shared values and common political institutions and principles of freedom, self-determination, justice, rule of law, equality and dignity of the individual (Hamilton, 1994). Though India is a low per capita income country, its large size, industrial base and technological sophistication make several deep integration issues highly salient. India may have led the developing countries' unsuccessful resistance to several issues in the Uruguay Round negotiations (Ahearn,1990), but since then, India has undergone a domestic policy revolution that has strongly affected its foreign economic policy (Haggard, 1995). The United States is India's largest trading partner with two way trade in 1997 totaling $10.9

billion, thereby crossing the 10 billion mark for the first time and reflecting an increase of one hundred percent since 1992. Since the introduction of economic liberalization policies in India, India's exports to the US since 1992 have grown at an average of 14.2 % per year in dollar terms. However, the level of bilateral trade is still only a very small fraction of USA's global trade. While US exports to India account for nearly 12% of India's non-oil imports and US is the destination of 18.9% of Indian exports, USA's trade turnover with India constitutes less than 1% of its global trade. In 1997 India's share in total US imports was 0.84%, an increase from its 1996 share of 0.78%. In terms of ranking, India stood 23rd among countries exporting to the US. India's share in total US exports was 0.53%. Considering the US Investment in India, the US is the single largest foreign investor in India accounting for around a quarter of all foreign direct investment approved from 1991 to September . One can also review certain aspects of the Indian defense technology scene to learn that India is a nation on the move.

The Defense Technology Scene

India is doing the things that should have been done 50 years ago and with so much of economic excitement being generated between the two countries, one should begin to expect mutual understanding and harmony in other areas of cooperation as well. Unfortunately in the defense technology area this is not completely true. What is happening is that the US is bracketing India with other countries that do not have a good record of non-proliferation. It is bracketing India with countries that are not democracies and with countries that have a history of blatant abuse of human rights.

However, India's defense establishment is at a turning point that will set its course for years to come. Its defense R&D is keen on new joint ventures for advanced technologies and is wide open for technology investment. So, Western defense firms looking for new business should turn their attention to India, where lies an attractive combination of a growing economy, a not so modern arsenal and a defense procurement process that is being revitalized for efficiency and value for the money. Such Western interest could herald a gradual reduction in its reliance on Soviet weapons, But, India has also started to put on a 'get-tough' approach with arms makers who think India will be happy with less than what other buyer nations receive in industrial incentives linked to the purchase of arms.

The Indian defense inventory is now comprised of about 75% Russian weapons and is bound to reduce over the next few years and will open the way for western arms makers provide there are no setbacks to the changes envisaged. Western dealers will have to remember that those who felt that tenacity and guile were required for success in the Indian market will now encounter a tough customer (Barnard,1995). Typical offsets for India have been 10 to 12 percent while other countries demand and get 100 to 200 percent. This equation will have to be rewritten since larger offsets and other industrial incentives are vital if India is to accomplish one of its chief national security objectives by making its defense industrial base self-reliant . India like other foreign buyers of US weapons and military planes is demanding more trade concessions and technology transfers in exchange and the demands are more closely tied to industrial policies. Since the mid

1980's foreign buyers increasingly have wanted commercial and military technology to broaden the purchaser's economy. American defense companies feel that without these offset agreements, they would lose the sales to other defense equipment providers from Russia, Europe or any other country (Cole & Cooper, 1996).

India's long dependency on the former Soviet Union for most of its weapons is now felt in many circles as a great deficiency that has to be corrected and that can only be done by identifying alternate sources of supply for building up India's own defense capabilities, which is not so easy a task. Interestingly, while India bought most of its weapons from Russia, 80 percent of the Indian students who study abroad choose America. Some Indians feel that there should be an "arms-length' relationship with any super-power. If India is reliant on any superpower for its weapons, this would leave the door wide open for coercive diplomacy. Despite the January 1995 visit of Defense Secretary Perry which was welcomed as an indication of American desire to improve relations, few in India believe the United states is willing to transfer to India the sophisticated technologies it needs to modernize its defense base. Despite the difficulties, Indian leaders believe the time is right to upgrade their manufacturing processes and technology. Defense companies all over the world are finding it an uphill task for business development. India, though on a shoestring budget, is willing to invest in carefully selected defense technology programs. The technology and industrial partnerships that India must have today are also to be found in the United States and Europe. However, it appears likely that American defense companies though keen on getting a foothold in India, do not have an easy time doing business with India for several reasons, the chief among them being the stringent but not so often (at least from the exporters point of view) sensible or reasonable US export restrictions.

US technology transfer restrictions to India are far tighter than those between the United States and NATO allies and of course one cannot expect it to be any different under existing circumstances. While there has been fairly extensive cooperation between the US and India, for example, on the Indian Light Combat Aircraft (LCA) Program, there are restrictions on the super computers, missile technology and even certain dual use items to India. On the other hand, many countries in Europe are willing to work with India in similar areas of technological requirements. This, from the Indian perspective makes the United States an unreliable supplier. In addition, for the transfer of technology to India, the United States has been bogged down by its need to make a choice between India and Pakistan. It is obvious that for the Indian subcontinent, conflict itself, or even the anticipation of it can be a powerful motivation for indigenous development of military technology and as Freeman states, new technology also spreads across national boundaries in response to military demands for national security. Efforts by the US or any other country to halt such transfer of technology may briefly retard it initially, but, in the absence of the removal of the sources of tension which motivate it, nothing can stop it for long (Freeman, 1994). The present state of affairs on the Indian subcontinent prove his point.

India has also engaged Israel in a dialogue regarding its participation in India's military modernization program according to a defense reports (Rodan,1996). India and Israel are also talking of furthering broad scale defense cooperation which will tie up weapon and technology transfer agreements. It is felt in Indian circles that these Indo-Israeli talks might create the possibility for technology transfer and service to service cooperation. However, Israel's close defense ties with China could be the first glitch in the cooperation. Added to this could be India's policy of maintaining close ties with the Arab world which may be at odds with Israel's drive against Islamic fundamentalism. Since US technology has found its way into Israel, the US might view India's cooperation with Israel as an indirect way of accessing US technology.

India seeks components and equipment in limited numbers for its exploratory research and development projects which are legitimate from the Indian point of view, but are viewed differently by the American side. Perhaps the alarm starts to ring that third world countries are beginning to acquire or are desirous of acquiring the first world's arsenal. Whenever the subject of technology and technology transfer is tabled, one comes across the stonewall of the Missile Technology Control regime (MTCR), the Non-proliferation treaty (NPT) and the Comprehensive Test Ban Treaty (CTBT). India is not a signatory to any of these regimes or treaties on paper even though her track record is clean for it is practiced. So when one views the apparently illogical way licenses are delayed or denied, Indian laboratories feel frustrated and look at Europe as a second source. But this availability of technology in Europe as a second source does not in any way improve Indo-US relations and in turn may be counter-productive, thus creating a vicious cycle. Since US Export Controls seems to be critical to the technology transfer problem, an in-depth analysis does seem necessary to investigate whether or not India and the US face a perhaps breakable impasse.

4
The Export Control Problem

The demand for defense technology, hardware or otherwise, by developing countries and the export or the export control or its denial by industrialized countries is an area that needs to be studied with patience by those who demand and those who control or refuse the transfer so that a clear view of the game and the game players is presented. Too many times have frayed nerves and frustrations been displayed across negotiation tables and this has given rise to loss of business, loss of project progress and loss of dignity in international relations. The exporting company, the controlling country and the importing country and the reasons that they get together must be studied to truly understand the export control dilemma. In the following analysis, it must be emphasized that it is only conventional components, equipment and weapons that are kept in view since no country will openly demand for nuclear weapons or equipment to build weapons of mass destruction.

According to Wallop and Codevilla, to believe in arms control is to accept that a nation possessed of serious reasons and weapons for fighting a war can and will set aside those reasons long enough to deprive itself of the weapons (Wallop & Codevilla, 1987). The real controllers of a nation's armaments are those people who have the power to decide what and how many weapons the nation's factories shall roll out. Those who started to advocate arms and technology control just after the last war were aware of the difference between actual controls and paper agreements. They knew that agreements were needed to ensure the practice of arms and technology controls. Today we are witnesses to the fact that several countries sign these agreements and then proceed to blatantly work contrary to all norms. It appears that agreements are signed more as a political gimmick. So the problem is that, as Morehead says, although the United States and other Western nations seek to use export controls to prevent the diversion of arms, military technology, and other strategic goods to the Soviet Union, Eastern Europe, and countries engaged in or supporting international terrorism, their enforcement efforts are confused, fragmented, and tend to lead to unnecessary conflict and discord (Morehead 1992).

Arms Transfers

It was a popular belief that the profits made by the arms manufacturers in arms trade must have been a contributing factor to the outbreak of the First World War. Since then, consistent attention has been paid to the arms trade because there is indeed a lot of profits to be made which of course has certain associated problems. The case of the Bofors weapons deal that rocked India in the early 90s unveiled alleged scandal and corruption that is still being investigated. Another cause for alarm is when suppliers find themselves facing weapons which they themselves sold to their new adversaries as happened to Britain in 1982 with regard to Argentina and to France in 1991 with Iraq. This consideration serves to stress the point that major military equipment must be exported only after the most careful consideration of the objectives and aspirations of the buyer. It must be realized that arms have a certain life span whereas political situations can change very quickly. Too often, practical considerations and risks which must be correctly identified are either played down when a deal has to be pushed through or blown out of proportion if the exporter has a hidden agenda.

Traditionally, in international relations, the military force of a country has gone hand in hand with its foreign relations. Today there are doubts expressed about the effectiveness and utility of this interaction. However, military force is still associated particularly with the ability of one nation to force another for a particular purpose and it is the technology that is available to a nation that can influence its capability during such show of force. The better the technology, the better the weapons and equipment and the better will be a country's performance in war provided that the operational capability of the defense personnel are fully exploited. So it should not be surprising that so much attention has been paid to defense technology and the arms trades between nations.

Since the capability to use force is derived from a nation's defense technology and weaponry, all such deals for transfer have to be closely monitored. Straight forward defense technology or arms exports requires a license from the exporting government. However as explained by Taylor (Taylor, 1994), under certain conditions, exporting agencies may also seek to illegally export the technology or equipment. The distinction between legal and illegal arms transfers can sometimes become blurred. In addition, since arms transfers can be suppressed or exaggerated, the evaluation of the nature of even legal conventional arms transfers is an extremely difficult job for the watch dogs of such transfers. Thus data on the arms trade have to be evaluated very carefully. Data apart and no matter whatever be the type of arms transfers, what is important is to understand why technology and arms exports appeal to governments, why governments import and the arms manufacturer who is the puppeteer in this game and at times can even become a sad puppet.

Appeal Of Arms Exports To Governments

Generally speaking, the face value of arms technology and arms exports is its commercial value. Arms deals can rake in enormous profits to the manufacturers. But the defense industry is not the only one at the helm. The armed forces also have an ax to grind for they stand to gain major incentives by voting for exports. The sales from arms help to reduce development costs for the armed forces during complete production as they benefit from the same kinds of economies of scale as the defense industry. So they can obtain more systems for the same budget outlay. Foreign sales also help them to fund the research and development costs for future generation systems. Certain aspects of the Russian and American philosophy on arms exports can be studied with perhaps more emphasis on the American one due to the aim and scope of this book.

When one considers the former Soviet Union, Gaddy (Gaddy, 1993) explains that the Russian factory overheads are lower than those elsewhere in the world; there is heavy subsidization of its defense output which encourages export and excessive militarization of the economy, all making a very conducive package for weapons exports and without too many inhibitions. In so doing, it is seeking partly to reduce political pressure for subsidies to industry and partly to provide enterprises with finance for their conversion efforts . One key aspect is that the governments with limited budgets can directly profit from arms sales through various fees and tariffs and governments are looking at export as a cheap way to preserve their defense industrial bases.

Now looking a little more deeply at the export policy of the United States, Goldring (Goldring,1996) says that while the United States debates domestic gun control on one hand, they are supplying enormous quantities of weapons around the world everything from shoulder-fired missiles to the best fighter aircraft. During February - April 1996, the US administration had announced its intention to sell sophisticated Stinger anti-aircraft missiles to Taiwan, aircraft and missiles to Pakistan, and high-tech fighter aircraft and other advanced weaponry to South America. Goldring calls these sales as short-sighted and counterproductive, threatening rather than serving US forces and foreign policy interests. In a news conference in Chile, US Defense Secretary Perry said that he wanted to loosen restrictions on sales of advanced weaponry to South America. The US administration has violated its own non-proliferation law and policy by defeating the Pressler amendment and endorsing the supply of advanced weaponry to Pakistan, adding to Indian anxiety. Though the US government has concluded that China's transfers of ring magnets to Pakistan violate US law, each time the US intelligence community comes up with apparently convincing evidence of Chinese weapon transfers, the White House buries its head in the sand (DN-1, 1996). Nevertheless, the administration proposes to loosen restrictions on transfers to Pakistan arguing that discussion of these proposed transfers preceded the violation of the law. Several members of Congress have questioned whether the United States should reward Pakistan's continuing development and refinement of nuclear weapons capability by supplying it with additional weaponry. Of course, the US is not alone in this interaction as China has been actively working together with Pakistan (IDR, 1988; DFAW,1990). The 'business of government being the government of business', most Western nations follow the US and Russia and are busy with their own export activities.

Members of Congress, analysts and the public alike had expressed concern about excessive weapons transfers to the unstable Balkan region. Yet, the US administration reportedly tried to push the bureaucracy to approve the controversial sale of Super Cobra helicopters before Turkish President Weyman Demirel's visit to the United States in late March 1996. Pressure for the sale came despite Turkey's human rights record, for Turkey's military supposedly uses US weapons to attack its own citizens (Goldring, 1996). This goes against the grain of the US administration's stand on human rights and is perhaps creating a threat to its own forces. No matter how one may view this, according to the forecast and analysis of the World Wide Conventional Arms Trade (WWCAT, 1994), while the nature of the political military issues that confront the US and friendly nations have changed, arms exports will continue to be a means of advancing US national security and foreign policy objectives.

One must also take cognizance of the fact that the 'Code of Conduct on arms transfers' requires the President to submit to Congress, once a year, a list of countries that meet certain criteria for eligibility to import American weapons. The eligibility criteria stresses on 1. a democratic form of government, 2. respect for basic human rights of citizens, 3. non-aggression against other states and 4. full participation in the UN Register of Conventional

Arms. These have been basic criteria in all US administrations. However according to Lumpe, 85% of US arms transfers during 1990-95 went to states which did not meet the Code's criteria (Lumpe,1998). This is in spite of Congress having the power to stop individual arms exports as per Arms Export Control Act, Section 36. Perhaps countries meeting all or at least most of the arms transfer criteria should be in a position to import US technology at least for the specific use of their own R&D. India for one is working on a great many R&D projects, basically to enhance its own defense technology capabilities. This could take years and does not in any way threaten western technological leadership or challenge western military power.

The motives for countries to export arms vary from economic, foreign policy and national strategic concerns and the weight of any single consideration clearly varies from case to case. Advanced military states have political and defense interests and allies have historically engaged in arms transfer and other technology transfer to support their individual ends (Morgan, 1991). The White House Conventional Arms Transfer Policy issued on 17 February 1995 gives a clear picture of how the US views the situation (WH, 1995). Economically, arms exports, like any other exports, should generate foreign exchange and employment. Employment consideration is important as many communities survive because of the arms trade. Another economic aspect of arms sales is that they can be used to recoup some of the R&D costs which were sunk into the system's development by the manufacturing state. These exports thus enable the state to pay a lower price for the equipment bought for its own armed forces.

To enhance the capability of friends and allies, major arms exporting countries must also invest in the training of personnel in these countries. This is made easier by selling equipment being currently used by its own defense forces. By the end of the 1970's the United States and other Western industrial countries, because of intense global competition, had begun to sell the third world the same types of weapons that they fielded with their own forces (Ferrari, 1988; Grimmett, 1990). To capture the market, barter arrangements are quite often used as payment. For example, in exchange for the MiG-29 combat aircraft sent to Malaysia, Moscow was willing to accept one-quarter of the payment in the form of palm oil. Export agreements with India similarly allow for payment to Moscow in the form of tea and rupees (O'Prey,1995).

Exports are also considered by countries to enhance their own military and financial security. Most technologically advanced nations have a major problem in sustained funding of their defense industry. So strategic considerations utilize arms exports to enable this sustenance. Since there is a tremendous change in defense technology every few years, the main consideration is that, although the rate of defense technological change can be very fast, major defense systems like combat aircraft, tanks and warships have a long service life. The defense forces prefer to update their inventories every few years with the latest equipment, thereby pleasing the industry. However once the service requirement is complete, what do factories and development units do to survive? The export of this equipment is the one thing that can keep these arms industries afloat.

So economically speaking, defense technology and equipment exports, like any other type of export, should generate local employment and foreign exchange. The actual economic benefits of a particular export depend on a variety of factors with the urgency of delivery to the recipient and the economic situation of the exporting country playing a lead role. The cost to the recipient, the delivery schedule and the sharing of benefits between the exporting firm and its government are then worked out to the satisfaction of all concerned.

Again, due to the slump in the arms industry and the end of the Cold War, the question of 'conversion' is found to be an expensive proposition as experienced by major weapons manufacturers who are now trying to trudge back to manufacturing commercial commodities. The shifting of the global demand for arms from the context of rivalry between the superpowers and their allies to providing for national defense in their own regional security needs has led to a decline in the total global demand for arms. But this does not mean that countries are not trying to acquire high technology and associated type of weapons. We are witnesses to increased violence and increased proliferation of defense technology irrespective of region. So even a goal of defense industrial conversion can be a justification for the export of weapons. Nations engaged in converting their excess defense industrial capacities to the civilian economy require dedicated resources. But as usual, resources are always scarce. This has prompted some governments and many defense firms to consider exports as a means to fund their conversion efforts. Discussing the systemic defense industry overcapacity in the United States and Russia, O'Prey (O'Prey, 1995) says that both countries must convert a large part of their defense sectors to civilian tasks and although the need for defense conversion is clear, the task has proved difficult to carry out. He adds that while empirical evidence indicates that diversification to relatively similar products within a commercial plant on an average is successful only half the time, the record is much worse in the defense industry. Thus the notion of simply redirecting existing defense production lines to new tasks is too simplistic and should, according to Bernstein (Bernstein, 1993) have a broader definition. Gansler (Gansler, 1994-A) advocates that a firm's existing engineering innovation, capital equipment, production labor forces, and skilled management are its principle assets and in most cases conversion success will depend upon how well they are utilized .

So there will be a lot of attention paid to defense conversion in the years ahead. While there is also a lot to be gained for countries to restructure their defense sectors by integrating defense and civil production (OTA,1994) as the gap between the two is increasing with the defense sector lagging behind, bureaucrats and technologists must recognize that conversion is a long-term process that may take years and they will surely have to shore up their resources even if they have to look towards arms exports.

The Arms Exporting Company

Even in the day to day interactions of the Indian R&D laboratories with the US defense industry, it is obvious that the decline of the international arms markets and the general decrease in defense spending by arms exporting nations, has led to intense competition for arms sales. While quite a few suppliers have merged or gone out of business, they are still competing for fewer and generally smaller sales which are so important to their financial credibility as the business base declines. Within this tight competitive environment, customers are demanding and obtaining better deals from suppliers. Financial problems such as low defense budgets in the recipient countries may also aggravate the situation for the supplier.

Many of the factors that motivate a government to export arms and technology also affect the companies concerned. As domestic demands reduce, most defense firms are starting to realize that they need to export to survive and that this trend is bound to continue into the future. While it is an accepted fact that companies need to sell in-order to generate a turnover and a profit, in the defense industry the competition is extremely stiff. In the United States and in Europe for example, the domestic markets are ruled by only a few major suppliers and therefore for the less privileged ones, international markets are the key to survival. These suppliers also acknowledge the fact that clinching the deal on a major piece of defense equipment with a developing country may take years of negotiations with huge expenditure as they need to go through government permission to transfer technology, field trials, license or co-production arrangements and price negotiations with subcontractors and the recipient. Finally the orders placed by developing countries may not be adequate to offset any drop in domestic requirements. In spite of all these headaches, export markets look quite attractive to companies starving for business. One issue that arouses curiosity is that there are certain exports that normally should be stopped by a country's export control enforcement, but for reasons not clear, are being permitted to continue with negligible control. A case in point is the US commerce department's promise to review regulations for US exports of devices used for torture, but crime control products may still be sent to NATO countries under license (Lelyveld, 1996).

Yet in spite of all this, the future of multinational export control remains in question today. The United States wanted to establish a side group of the six major arms exporters - Britain, France, Germany, Italy, Russia and the United States - that could meet to deal with crises and to coordinate export policies (Erlich, 1996-A). The future of this arrangement still remains undecided, the pact itself does not control the flow of arms, but only provides a forum for the exchange of information on arms shipments. One must in the same vein consider White House and US Commerce Department officials resistance to the calls to tighten the reins on supercomputer exports to China (Opall,1997). The US lawmakers and nonproliferation advocates have campaigned for government supervision of all supercomputer sales to most destinations on the grounds of national security. The administration has put the onus on individual exporters and held them principally responsible for weeding out the bad from the good customers. William Reinsch, undersecretary of commerce for export administration told a House National Security military procurement subcommittee panel April 15 1997 that 1,100 high-performance computers, valued at more than $550 million, were exported from the United States from January 1996 through March 1997. Of those, 46 were exported to China. One should remember that this is more computing power than is owned by the Pentagon and all the national laboratories put together according to Stephen Bryen, a former Pentagon official who founded the Defense Technology Security Administration (DTSA). So over the years, there has been a see-saw of decisions by the administration, for till recently US government policy required export licenses only to countries that posed security or nonproliferation concerns as per US definition. In other cases, industry could sell computers capable of computing 10,000 MTOPS without any government supervision. So, till recently, this left the field open for countries like China to purchase computers

In Defense Of My Country

up to a 7,000 MTOPS level. On the other hand, India which already has visual Silicon Graphics Systems for its flight simulation and is looking for upgrades to existing systems, is denied the export of these systems for the simple reason that they are just above the 2000 MTOPs level which is the present cut-off criteria.

However, even though the US acknowledges her political-military changes, the changes in the economics of the defense industry, the decline in the DOD requirements, the need for technology and arms exports for the financial health of the defense industry, the impact of these export sales on national security and the need for government support to the industry to export, the US has problems in maintaining a consistent export policy. One hears quite often of a deadlock between the White House and Congress over export controls with no one being able to devise a compromise solution that could make its way into law. William Reinsch agrees that some of the initiatives proposed by the Aerospace Industries Association (AIA) deserves careful study. Some of the points as mentioned in the 9 June 1997 issue of Defense News include : 1. Issuance of program export licenses for a product and all subsequent shipments of spare parts as long as the basic capability of the product is not changed; 2. Elimination of licenses for products destined for countries that maintain export restrictions similar to the United States; 3. Certification of company employees to issue certain export licenses on behalf of the federal government; 4. Permission for US companies to honor existing contracts, except when multilateral sanctions are imposed to nullify all contracts; 5. Permission for companies to support products previously exported to a country, even if new sales are prohibited. These practical suggestions apply very well to the

Indian dilemma of importing from the US as it is of real life significance for all the Indian DRDO labs and establishments.

Countries like the US could help developing countries create strong economies and sound technological bases for societal goals before selling or even attempting to sell them full-fledged military equipment. Today, quite often the US promotes sales of weapons and in the same breath indirectly limits export opportunities for other American industries. This can be explained by the fact that a country like India needs to buy components and limited equipment to develop its own technological programs. These components and subsystems come from the small US defense industries. If they are permitted to export these commodities, it would not only help these industries but also enhance India's technological skills in putting together its own systems. India could thus be a long term customer or perhaps a partner to the US. Refusal to export could drive India into the arms of countries willing to export or force India to develop her own defense base independent of the US, thereby defeating what the US intended all the time.

Arms Importing Governments

Defense procurement is a problematic task for all governments in developing countries. Interactions with personnel involved in the process confirm that no government can claim to be a model of excellence in this respect as there are too many variables involved. The technology available is so much and so varied but budgets are limited and developing countries find it extremely difficult to keep up with the state of the art technology. Some countries simply buy equipment "off the shelf", but they too must do a fair bit of homework before the procurement. Equipment is sometimes purchased by countries after recommendations by self-centered high ranking individuals who choose the particular equipment because of the personal benefit involved. This is also very true about recommendations from responsible operational staff who wish to go overseas for training and testing when their own countries do have the state of the art equipment and testing facilities. Another aspect to be considered is that armed forces everywhere give a lot of weightage to prestige and may recommend the purchase of equipment that will enhance their prestige and that of the country.

Whatever be the reason for the import and even if the need for export is pressing, in all fairness one must acknowledge the fact that the economic needs of exporting countries do not drive arms transfers as much as the requirements of the purchasing countries. The 28 July 1997 issue of Defense News mentions the Aerospace Industries Association as saying that Asia has supplanted Europe as the top market for US aerospace exports and that the sales covered civilian and military products. Aerospace sales in Asia accounted for about 13 percent of total industry sales and 41 percent of all exports, with sales to the European Union comprising 39.7 percent of all exports. It is also widely felt in US industry circles that the Asian market is critical to the US industry's success and restrictions on defense sales to these countries will turn them towards Russia.

The combination of regional problems and security issues with economic success and the need to establish ones defense capability is the ideal environment to attract arms and technology exports. South Asia is a good example of regional instability and fear with a fair amount of resources that dictate the need for imports by those countries. India's motivation to acquire arms to match China will always alarm Pakistan and Pakistan's acquisitions are a concern to India. But one should not forget the fact that the success of arms trade is to some extent dependent on the countries involved. Countries that import technology and equipment go through a cycle of absorbing the technology, utilizing the technology, implementing the technology into their own R&D efforts before they think of acquiring further technology. The period of this cycle depends on the infrastructure and trained manpower in the country.

For a developing country, investing in defense technology and defense equipment is a drain on precious resources, resources that could be well utilized elsewhere. In an attempt to create security through military strength, investment of these resources in dead capital like defense equipment may provoke internal problems. This reduced investment in the national programs can create unemployment and poverty. Apart from protecting the national security of a country, the only other saving feature could be the spin-offs that defense technology might have for the civilian population.

Influence Of Exports On Foreign Policy

The import and export of defense technology and defense equipment is frequently used as a foreign policy trump card to influence the attitudes of either the importer or the exporter or as a token of thanks for help given or as a reward for 'good behavior'. There is a limit to the influence an exporting country can wield over a recipient country for it is still a buyers' market and importing countries like to limit the extent to which they commit themselves in any way. India for one often tries to procure from a range of suppliers. This tends to reduce its dependency on any one country and thereby avoids being held hostage for spares and delivery schedules at the suppliers whim. Again, importing countries also try to build their own conventional backup industry in which case arms imports gradually lose sway. For the exporter, arms export policy is deemed as a failure when the exporter faces his own technology and equipment in the hands of an enemy. So exporters must exercise caution when major defense equipment is sold and must do so only after the most careful examination of the purposes and aspirations of the recipient. However, practical considerations and risks must also be evaluated side by side. To refuse to sell arms to a country which feels the need for its legitimate self-defense or for the development of its own technology base may be construed as an unfriendly act, thereby creating a feeling of hostility apart from other issues.

The importance of defense technology and arms has a direct bearing on foreign policy. To underline this thought, Cardomone (Cardomone, 1995) explains such weapons and technologies, in the wrong hands, could have the power to change military balances, disrupt military operations, and cause significant casualties to defense personnel. In the US, the spread of conventional arms is the engine that drives a whole host of foreign and domestic policy decisions. Boyer (Boyer, 1994) states that arms control has now become part of preventive diplomacy and can no longer be limited to a small number of players. Its main focus has accordingly shifted from nuclear issues towards biological, chemical and, particularly, conventional weapons. There could be a host of new problems because of the changing nature of technology. Unfortunately, the US does not always follow her own warning and presents justifications for export that advocate 1. weapons exports promote US foreign policy objectives, 2. Arms sales enhance US influence with foreign governments and 3. Arms sales allow inter-operability with foreign forces. So, for the US, arms sales are still seen as a legitimate and critical tool in formulating foreign policy.

Decline of the Arms Market

The value of the global arms exports to the third world has declined from 1986 till date basically because of the end of the cold war and breakup of the Soviet Union - the same factors that had started the slide of the reduction in domestic demand for weapons in Russia and the United States. Without involvement in statistics, one can safely say the over capacity of weapons and defense production capabilities was caused by the defense buildups of the 1980s for the reasons given (Velocci,1994; Morocco,1994). O'Prey states that most countries neither confront the threats to their security nor possess the economic means for maintaining their previous levels of militarization (O'Prey, 1995). There are some former large importers like India, who now prefer to try and develop and protect their own defense industries rather than buy everything from abroad. It is also a fact that those countries that can afford to pay hard currency for weapons already possess more arms than they need or want, while those that crave western technology and arms do not have the ready resources to pay for them. So as Durch explains, the next best option available for these countries other than purchase expensive systems is to consider upgrades to their existing defense systems. These upgrade programs can extend the life and increase the combat capabilities of existing aircraft, combat vehicles, warships, and missiles and are also less expensive (Durch, 1994). One also comes across cases where military powers are willing to transfer surplus equipment at ridiculously low prices and sometimes even for the goodwill ! Durch also elaborates that there is a glut of reasonably priced older but still effective weapons available because of the various arms control agreements, continuous modernization programs and defense downsizing. Developing countries would also grab at this option rather than invest in the more modern technology. In addition, Finnegan (Finnegan,1994) mentions that traditional arms exporters now face new competition for supply from developing producers such as India, Taiwan, and Brazil, even though it is clear that India has not made it to the aggressive supplier list for a number of reasons. But new exporters have entered the market at the low and medium segments of the technological spectrum and are also contributing to the increasingly competitive nature of the upgrades market. Another aspect that is striking is that arms importers have become much more demanding for quality equipment, insisting on technology transfers, co-production agreements, offsets, countertrade requirements, and after-sale service. They are no longer gullible enough to purchase second-rate systems from the major powers.

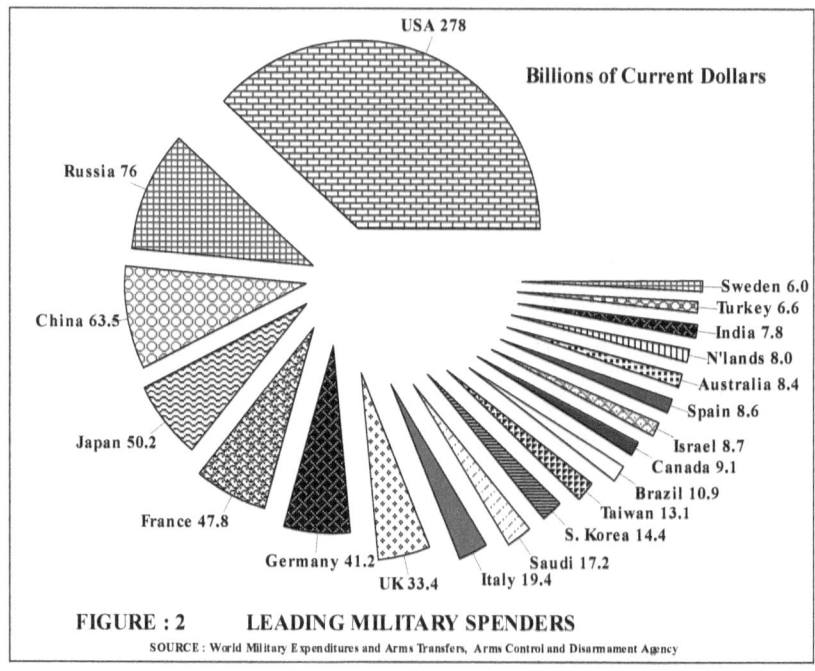

FIGURE : 2   LEADING MILITARY SPENDERS
SOURCE: World Military Expenditures and Arms Transfers, Arms Control and Disarmament Agency

Arms Control and Disarmament Agency (ACDA) information (Figure 2) shows that states in the developing world and in the world as a whole are tending to devote less of their defense budgets to arms imports, perhaps because of increased domestic arms production. Taylor tracks the arms imports during the last decade and says that the average annual world arms imports in the 1986-9 period were up only 8 per cent on their 1979 level, whereas world military expenditure was up by 24 per cent. He adds that among the developing countries, annual arms imports were down by 2 per cent in the 1989 period compared with 1979, while military expenditure was up by only 9 per cent (Taylor, 1994). The 1996 World Military Expenditures and Arms Transfers shows that as a group, the developed countries have had a 7% annual decline since 1992.

In Defense Of My Country

The SIPRI 1997 yearbook describes the current NATO military spending as continuing to decline in 1996. As NATO expenditure is the dominant component of overall world military expenditure, the yearbook says that the decline in aggregate global security expenditure noted in recent years was maintained in 1996. Malaysia, Singapore and Thailand maintained the Southeast Asia's reputation as the fastest-growing defense spender in 1996. Looking at the Indian subcontinent, defense spending has reached a plateau even in 1996. While the lack of growth in official Indian defense spending in real terms seems to skew the overall expenditure picture, the yearbook specifies that in Pakistan the military expenditure grew in real terms by some 2% and by a 29% in Sri Lanka during the same period.

Any analysis aimed at explaining the arms trade should not neglect the difficulties which a recipient faces in trying to decide what should be acquired, and the consequent criticisms, which are often leveled against individual selections. Moreover, many developing countries lack the technical expertise to evaluate the relative merits of the different systems being offered even if they do come in for field trials. They sometimes lack the capability to evaluate all the documentation involved which can be quite complex for any modern defense system. For the purchasing government, arms importing issues are only one element in the broader problems of defense acquisition and procurement since there are a host of problems inherent within each system of government, which accompany any acquisition. Armed forces everywhere place a heavy stress on symbolism and therefore it is possible that they will often be attracted to equipment which they see as adding prestige to the services and the country. In some countries it is difficult to ensure that those responsible for acquisition will be unbiased and opt for the best systems rather than those that bring in the most personal benefit.

With all the push and pull of arms sales and the strategic and financial attractions, developing countries find it difficult to comprehend all the effort by Western nations towards arms and technology controls when in fact there is a genuine desire by them to sell arms and technology and a genuine desire by importing countries to get their hands on the technology and equipment. A study of the US perspective and its fears in this regard will bring to light many of the problems and issues that may not be evident at first sight to a third world nation in need of the arms and technology. The effectiveness of this study may be enhanced by first studying the Indian perspectives and requirements.

In Defense Of My Country

5
Indian Perspectives

William Rees-Mogg says that anyone who wants to understand the modern world must make a personal passage to India – a nation which has the deepest and most resilient culture of the four likely economic super powers of the next century. India is more stable and politically advanced than China and not yet denatured by the modernism of the United States and Europe. He adds that 'Indian civilization is a great lake into which the rivers of different cultures have flowed for more than two and a half millennia, each depositing a new layer' (Rees-Mogg, 1996).

In India one can see the inevitability of Asian economic expansion (Garten, 1994), despite the serious problem of population growth. The high growth of Asian economies is largely based on this transfer of technology, which provides outstanding investment opportunities. Rees-Mogg sums up Indian capability by saying that given the same software, an Indian keyboard operator is as or much more productive in comparison to an American or Japanese. The world ought to know that the keyboards are coming into India and India has a middle class, as large as the whole population of the United States, capable of using the keyboards effectively.

Till date, Indians have felt prouder of their religious and cultural tradition than of their economic future. But today this spirituality has made an interesting congruence with modern thought and ambition. Everett M. Dirksen once said that, "There is no force so powerful as an idea whose time has come." In the global scenario, India's time has come. Technology advances will ensure the qualitative edge India needs and the time is ripe for the Indian technical capability to be turned into a national resource.

General James R. Clapper Jr. in his prepared statement to the Senate Armed Services Committee describes much of the Third World as resting on a bed of kindling wood with unpredictable flash points. With reference to India he mentions that India and Pakistan remain a concern because of the presence of very large forces in close proximity across the line of contact, as well as their pursuit of ballistic missiles and weapons of mass destruction. Islamabad and Delhi are believed to be preoccupied with internal problems and recognize that war is not in the interest of either. However, as always, this remains a potential flash point because of the danger of miscalculation and the prospect for rapid escalation of a crisis (Clapper, 1995). This must prompt us then to try and understand India in her global setting, her security strategy and why she behaves the way she does sometimes even in the face of global opposition.

Indian Neighborhood

All over the world, history and international relations are hinged on the rapport between neighboring nations and the influence each has on the other. India in its geographical setting is no exception and her relations with her neighbors should be viewed in this perspective (Damodaran, 1986). One must concur with Snyder when he states that many of India's problems and challenges are those inherent in her geographical and historical situation and also because the United States, Soviet Union and China have involved India and Pakistan in their own rivalries, as well as they themselves entering into Indo-Pakistani disputes (Snyder, 1995). As the largest democracy and most powerful state in the region, India cannot escape her destiny and avoid shouldering the responsibility of keeping the peace in the region. However, almost all countries in the world have problems to varying degrees of seriousness with their neighbors and India is no exception.

All the countries in South Asia face the associated problems that come with development and modernization. While there has been some talk of fear of India because of its size, its economic and military pre-eminence, one factor often missed is that such fears could often arise out of one's own diffidence. If viewed very carefully, one will see that India is a functioning democracy and has managed to deal with its linguistic, religious, orthodoxy and other social problems to a greater degree of maturity than some of her neighbors and may not have the time or the inclination for devious intentions. This has been true over the years and Indian leadership was 'wise in not focusing too much attention on India's neighbors and being more interested in the game of nations at the international level' (Subramanayam, 1986).

Policies and developments in the neighboring states have had a profound impact on India's own development and security. Because of the conflicts with China and Pakistan, the effect of the Cold War in the region, and several domestic issues, India could not develop her relations in the sixties and seventies. To avoid being too disheartened, Jha compares India's predicament to the political preconditions that were fulfilled by the European

Economic Community countries, before they entered into their own collaboration (Jha, 1986). With a more confident India opening its doors and seeking to re-engage itself actively with the region, a new chapter is beginning in Asian relations (Dutta, 1992).

Afghanistan, Bangladesh, Bhutan, China, Pakistan, Maldives, Myanmar, Nepal and Sri Lanka make up the immediate Indian neighborhood . India shares a 3325 km border with Pakistan, 3439 km with Tibet and China, 1751 km with Nepal, 4439 km with Bangladesh and a coastline which is 7683 kilometers. The calculated way in which the British Empire wound up its business in India, as they have done elsewhere, kick started the turmoil in the region. The British legacy of the first partition soon led to an inevitable second one and as a matter of fact, most of India's diplomatic preoccupation after independence can be traced to her territorial disputes with Pakistan and China. This territorial preoccupation is justified if one views it in the framework of the region and not merely as the dissatisfied bickering of unfriendly neighbors.

The China Factor

The Himalayan mountain range separates China from the subcontinent . So China could be considered both as a regional and as an extra-regional power even if it is physically a neighboring power of five of the countries in South Asia . But the Chinese have never concealed the special interest they have in the fortunes of India's smaller neighbors. It is also true that in all disputes and divergence between India and her neighbors, China has sought to be on the side of the other country in a fairly visible manner. China's aggression into Indian territory was to prove her strength and power more than trying to set right the border dispute. While this cannot be repeated so easily today, things that worry India about China are the threat posed by China's missiles, transfer of missiles to Pakistan and her growing nuclear ambition. The May, October and November 1995 issues of the Risk Report cover these aspects extensively (Risk Report, 1995). What is baffling is that the US does not seem to take steps towards sanctions against China even with irrefutable evidence (Smith R.J., 1996). Also intriguing is that China has been engaged in a massive effort to acquire defense technology in the United States (AMR, 1996). Of late, rumors have surfaced in Washington DC that the US administration has decided to offer China sophisticated missile technology on certain conditions (Haniffa, 1998). Again Bill Gertz in the 14 April edition of the Washington Times describes the current confusion enveloping the passing of missile development information from two US companies to China (Gertz,1998-A). These developments are bound to perturb India's new government. What the US has to consider is that, if not today, India could in course of time, be a real balancing factor to China in Asia and it will therefore seek checks on India, as India must on China. It has been a major characteristic of and flaw in US policy to regard those not wholly with it as against it (Bajpai, 1986). Washington should know that India may be far behind China in certain areas today, but with the way India is growing, it is projected to give China a run for her money (Haniffa, 1996-A). China maintains a strong security relationship with Islamabad and as long as China has unresolved territorial disputes with India, it is bound to continue to regard Pakistan as an important ally. Thus, while Chinese arms sales to the countries of the Middle East and Persian Gulf can best be explained in terms of economic motives, Chinese military cooperation with Pakistan has political overtones which, according to most Indians, bodes no good for India. Washington has been trying for 15 years to interrupt the flow of Chinese nuclear technology to Pakistan, acting partly out of opposition to any nuclear proliferation and partly out of a specific anxiety that persistent tensions between Pakistan and India over Kashmir and other matters make South Asia the most likely place in the world for a nuclear conflict (Smith R.J., 1996).

From an Indian perspective, the silence in Western circles concerning the Chinese nuclear weapons program and its impact on regional and international security is politically motivated. In 1971, when the United States decided that a rapprochement with China would be a strategic benefit, much of the rhetoric decrying the dangers of Chinese nuclear weapons disappeared. The US Defense Department, which had routinely used the Chinese nuclear program as a benchmark for sizing US nuclear forces, ceased to emphasize the dangers of the Chinese threat. One must realize that during this same period China systematically improved her intercontinental missile capabilities and poses a far greater threat now than at any time in the past. Indeed, in view of the alarm sounded in the US Congress in 1989 and the heated debates that followed, when India first tested the Agni missile, one would have been more or less convinced that India, rather than China, posed the greater threat to the United States. China, despite various official statements, has not agreed to abide by MTCR guidelines and appears to be continuing its policy of missile sales to the Third World, in spite of repeated pressure from the US administration to halt such sales (Erlich, 1996-B).

The US must now sit up and take cognizance of the threat that they face from China if one goes by the CIA report which states that 13 of China's 18 long range strategic missiles have single nuclear warheads aimed at US cities (Gertz,1998-B). This only proves that China views the United States as its major adversary and undermines the US administration's statement that there are no missiles aimed at her cities. It will also tend to undermine the thrust of some US business houses that are strongly lobbying to transfer nuclear technology to China and perhaps the invitations from the Department of Commerce to US firms to reapply for licenses (Opall,1998) once denied will have to be rethought.

China is using its economic growth to advance its military modernization program by procuring sophisticated Western weaponry and advanced military technology. Fisher (Fisher,1997) explains how several friends and allies of the United States, including Russia and Israel, are selling advanced weaponry and military technology to China, and several European countries, among them France and Britain, also are interested in getting into this market. India views this as a dangerous strategic development with associated repercussions for the region and the rest of the world and is concerned with its own security. There seems to be only a halfhearted attempt by the Clinton Administration, to at least slow down China's ambitions. Countries like Russia, Israel and European nations who are major suppliers to China are not interested in the long term effect as the attraction of supporting their own defense industries is high. The US administration's attitude toward supply of nuclear technology to China is an added grief.

In spite of the existing border problem, there is a certain mature and deliberate withdrawal and shelving of the problem by India and China, but as long as China remains an accepted nuclear power, India will not give up its nuclear ambitions. As far as India is concerned, she may take shelter under the knowledge and understanding that strategic deterrence may be the rationale for acquiring missile capability (Harvey, 1993). The Pakistani bomb may provide the political fuel for pressures on the Indian government to continue her programs. Fundamentally, however, India regards herself as a competitor with China, not with Pakistan. Pakistan is a dangerous irritant, but China is a regional superpower.

Since China exploded a nuclear device in 1964, India has refused to participate in discussions on nuclear non-proliferation on the grounds that nuclear disarmament must be comprehensive, not selective by region or state. India will not allow itself to be put into the category of a Third World state while China receives honorary superpower status based on its weapons technology. India's preoccupation with Chinese nuclear capabilities goes far beyond the mere performance characteristics of her weapons. Until the political dimensions of the proliferation problem are taken into account, regional arms control will remain elusive (Kemp, 1992).

What is also disturbing to India is that the US seems satisfied with China's nuclear controls (Morse, 1997). US senior State Department officials have declared that China's concrete actions as well as authoritative assurances regarding its controls on nuclear technology and hardware that proved sufficient for the Clinton Administration to implement the 1985 U.S.-China Agreement for Peaceful Nuclear Cooperation. The decision makes possible U.S. sales -- potentially worth billions of dollars -- to China of equipment and technology for 'peaceful nuclear programs'. All this based on pledges that China has made and has violated over the years. Erlich (Erlich,1997) mentions that China has sold weapon technology to Pakistan and Iran at least nine times since 1995, according to unclassified data compiled by the US administration. Each of these cases had the potential to violate U.S. laws, thereby inviting sanctions. However in each case U.S. State Department officials steered clear of disciplining China, adopting instead a policy of engagement. In all these cases, congress has accused the administration of very narrowly interpreting what was thought of as very specific legislation. According to the committee's count, in addition to the cruise missiles and chemical weapons technology, China has sold to Iran missile guidance systems and computerized machine tools, gyroscopes and other missile technology. Pakistan has received high-technology equipment for use in a nuclear facility, blueprints and equipment to manufacture nuclear-capable M-11 ballistic missiles, M-11 missiles or components, ring magnets for uranium enrichment, revealed in February 1996. Evidence of a sale of M-11 missiles or components in December 1992 triggered sanctions from August 1993 to November 1994 but there have been no penalties since then . The US feels that China's compliance with U.S. policy and international norms is increasing, but now China is Iran's largest arms supplier since Russia joined the Wassenaar Arrangement. Indians tend to be cautious and going by Chong-Pin Lin (Lin, 1988) who feels that ' it is not totally accurate to state that China has been a defensive rather than an expansionist nation' judging by the growth of China's territory over the centuries, there is enough justification for fear in the subcontinent.

R. J. Augustus

The Pakistan Factor

Pakistan as a major Islamic country is looked upon by other Islamic states as the stalwart of Islamic ideology. There are others that try to kindle the flame of Islamic fundamentalism that could become an insistent and very relevant problem. These ideological differences could have significant consequences. Yet in spite of the three wars India has fought with Pakistan and the major border conflict with China, the major game players have held on to their sanity for the relationships with India and her neighbors have had many unique characteristics. Even the wars were well controlled, in the sense that the civilian populations were not targeted. But there could be a certain amount of apprehension with Pakistan's top nuclear scientist Abdul Qadeer Khan contending that Pakistan's missiles were capable of targeting every Indian city in case of war and that his country's nuclear program was never capped or ever rolled back (Najeeb, 1997). For India, the unfolding drama of the Ghauri missile test and Moscow's reconsideration to sell arms to Pakistan (Radyuhin,1998) to improve relations are other causes for worry.

Perhaps most of these problems could be easily resolved, if there were not some extra-regional interests who see these minor disputes as a means to keep India in line in a world of increasingly intense economic and strategic competition. This can aggravate certain situations or disrupt existing cooperation between the countries. Both Moscow and Washington have a special interest in the region because of their own strategic calculations and with their influence are also capable of reducing the hostility in the region. It is but natural that each would try to influence India and her smaller neighbors by every means at their disposal. Another problem is that all the countries of South Asia have large emigrant communities in both the developed and developing parts of the world and this could make Indian society vulnerable to pressure (Abrams,1995). As of today, however, this is only a part of a global phenomenon. It could become more important to India if there was a repetition of the same phenomenon in the subcontinent or if these intra-regional developments could be exploited by interested foreign powers.

However more optimistically, it is possible to solve the difficulties that India has today with her neighbors but this calls for rational discussions and detailed negotiations in an atmosphere free of tension and one-upmanship. It is fortunate that this is not as impossible a task as it appears because the states involved have shared the same traditions and legacies for several centuries even during most difficult moments of their tumultuous history. This is an important angle to the regional issues that most western nations overlook. In South Asia, the most effective way of combating unfriendly elements would be to strengthen regional solidarity and cooperation. In the meantime the subcontinent is as far away from the creation of a zone of peace in the Indian Ocean as when it started. It takes time to get rid of old animosities and biases. India will have to learn to live with this for many years to come and is not an easy task for the ruling governments. India's policy towards her neighbors has common objectives that include establishment of friendly relations, cooperation wherever it is possible in economic and other matters and arriving at a consensus on common security perceptions, preventing non-regional powers from undue meddling. However, as explained by Bajpai the prescriptions for India's policy towards each country must necessarily be different and without sentiments clouding her judgment, keeping in view India's national and security concerns.

Indian Security Strategy

Today India is on her own for the first time in centuries and is undergoing a poignant and apprehensive transition between her traditional culture and the practices and influences of the industrially developed world. One can stand in almost any city in India and observe the simultaneous existence of many centuries of Indian history – a heady mix of the past and the present sometimes being rudely brushed aside by straws from the future. India has stood the test of time even though she threatens to be engulfed by her ethnic, linguistic, religious, caste, and internal regional rivalries, on a scale most Americans will find difficult to imagine. If one imagines the whole of Europe under one government, it might come close. This cauldron of inherent concerns seems at times to prevail over national concerns and to threaten India's fragile coherence and national integration. At the same time, according to George Tanham (Tanham,1992), most thinking and decision making Indians appear consumed by personal and regional competition for political power. This short term focus tends to diminish India's claim to greatness and ambition to play a larger role on the international stage. The extraordinary Indian diversity spread over the country is now being fueled by the urge to change with the rest of the world. There are rapid economic, technological, and social innovations changing the visage of Indian society and its way of thinking. Yet, at times, perhaps because of the 'misused freedom' of democracy critical issues are not achieved and ambitious plans are uneven and often discontinuous. Indians have started to realize that size, population and culture alone cannot fill her sails and lead her to international recognition. So now as the new government in India takes charge, Banerjee writes that they face a 'daunting task of dealing with a void in strategic thinking, an outmoded defense structure and a severe morale problem in the Armed Forces' (Banerjee,1998).

Indians have always considered the northern mountains and the surrounding southern seas as protective barriers against outside interference to the peninsular, but history through numerous military invasions has proved otherwise. With a thousand years of foreign rule, India has been able to accommodate these invasions in various ways, giving rise to today's diverse and modern India. While Indians have been content to stay inside the subcontinent with little interest in the outside world till recent times, her emerging economic potential and military capability are beginning to change their thinking. India has discarded the blinkers of sub continental foreign policy and aspires to be regarded at the global level on par with China. Also changing is India's view of defense strategy and realizes that power is an important variable in the international political arena and that there should be a translation of potential power into actual power which is so important from the point of power predictions (Baldwin, 1983). With a new foreign and defense policy on the anvil it is hoped that the Indian government realizes that security strategy is very different from bygone days since threats today are varied. Perhaps this realization has prompted them to revive the idea of a National Security Council (NSC) to help guide post-cold war policy and present alternative ideas to government leaders from outside India's entrenched bureaucracy (Raghuvanshi, 1996-A). This idea which was resisted by the Indian bureaucracy earlier is definitely needed to lift India up from the quagmire of old cold war thinking and to create a long term defense and security policy rather than today's reactive approach to any issue. This change in Indian thinking according to Dixit (Dixit,1998) could be signaled by an increase in defense spending and greater faith in indigenous research and development leading to a bigger thrust in defense exports and defense cooperation.

The nuclear issue too has dogged India over the years and it is felt that China achieved great power status, including a permanent seat and veto power in the UN Security Council, by acquiring nuclear weapons prior to the cut-off date inscribed in the Nuclear Non-Proliferation Treaty. India, which has not tested or deployed nuclear weapons since its 1974 test, has received no such status even though several congressmen favor the idea (Arora,1998). Indians feel that only by acquiring nuclear weapons, will they be taken seriously as a great power in their own right. For India's nuclear capability to be equated with that of Pakistan, and for India to be subject to the same superpower reprimands as Pakistan for its high technology achievements, only adds insult to the injury of not being taken seriously (Kemp, 1992).

Of course, no Indian security strategy will succeed if the Indian economy does not become more potent, stable and efficiently managed. This challenge to economic development and governmental effectiveness is the greatest threat to Indian security. Another realization is that military hardware alone is not the key to power, status and security for this could lead to inadequate economic development, isolation from the global economy, technology embargoes and increased social and political restiveness. Perkovich (Perkovich, 1995) says that India's interest would be much better served when the full potential of India's possibilities in Asia would be realized through further progress on the road towards economic liberalization, the revitalization of India's federal polity; India's ability to liberate itself from the current quagmire of Indo-Pak relations, a faster pace of Sino-Indian normalization and the introduction of significant political and strategic content in India's relations with the United States and Japan.

Regional Relations

Unlike the United States that is flanked by Canada to the North and Mexico to the South, Indians feel that they are in a bad neighborhood. Several neighboring countries have nuclear weapons or are capable of putting them together at short notice. So India has to be prepared for any eventuality and must be prepared to deal with any new security challenges. India acknowledges that a force structure that is not maintained in a state of readiness appropriate to its mission will be ineffective. India remains preoccupied with potential threats from China, and as an emerging world power, chafes under the nuclear double standard propounded by the established nuclear club. Pakistan already suffers from a significant inferiority in its conventional military forces vis-à-vis India, its principal strategic concern, and looks to China for nuclear assistance to help keep the field level in case there is a bully off with India. This triangular relationship complicates traditional arms control and technology transfer solutions. Into this cauldron one must mix the political ingredient that the Indian and Pakistani prime ministers are being accused of talking tough but acting weak. Indians accuse the US of spending two decades improving its relationship with China, often at the expense of other countries, including India and so they have often made it clear that they considered US efforts to meddle with Indian security affairs and military aspirations as impertinent and patronizing (WT, 1989). Again, they argue that the US should consider the region as a whole inclusive of China, India and other states. A dialogue at that stage would more likely lead to some specific areas of profound Indo-U.S. cooperation and such a partnership would not be an axis against China, but would facilitate normalization in South Asia. However, events such as the Indian Agni missile demonstrator launch and the US-Japanese agreement (Heinz, 1989) have highlighted the deteriorating ability of the United States to dictate or even to influence significantly the industrial and security policies of emerging regional powers (Nolan, 1991). The recent Pakistani launch of the Ghauri missile underscores the point.

Though India's goal should be the transformation of both China and Pakistan into benign neighbors, the humiliation of the 1962 war, the unresolved border problem and China's influence in Pakistan remain hurdles to any normalization of relations with either country. India has tried to increase the dialogue and to establish confidence-building measures with China and although détente has been strengthened between them, a number of difficult issues in this complex relationship remain unresolved. China carries clout at the regional and global level and Deshpande maintains that the Chinese nuclear explosion in 1964 began the process of the recognition of China and Nixon's visit to Beijing in 1972 completed it (Deshpande, 1986). The general consensus among a few of the Chinese and Indian scholars at a 1996 symposium held at Marquette University on 'Comparative Reforms in China and India' was that there is no comparison.

In Defense Of My Country

Indian defense and security analysts become peeved when Washington policy analysts and other strategic thinkers talk of a balance in the region between India and Pakistan, which Delhi believes is a misplaced concept, because the balance of power in its eyes is vis-à-vis China and not Pakistan. (Haniffa, 1996-B).

India possesses a sophisticated indigenous defense production capability (Karp, 1988) which sets it apart as one of the few nations in the Third World capable of acquiring, licensing, and developing state-of-the-art weapons systems. Kemp states that a pro-export policy was announced in February, 1989 when the Indian government decided to increase exports of its domestically produced weapons to finance imports of high technology weapons and upgrades (Kemp, 1992). What really matters from the technological and strategic point of view is the defense preparedness of China and Pakistan vis-à-vis India. A look at how relevant countries in the region spend their defense budgets, with reference to their size, would bring some clarity into this gray area.

Defense Spending Among South Asian Neighbors

Experts are of the opinion that China's thinking is more far-reaching and pragmatic compared to India's as far as defense technology development and military modernization is concerned. Giri Deshingar, director of the Institute of Chinese Studies, Center for the Study of Developing Studies, in New Delhi, says that India's armed forces, from now on can be quickly modernized only on the basis of foreign purchases and not on their own. In contrast to this, when the Chinese People's Liberation Army modernizes its hardware, it is going to be largely indigenous and much better than what they have today. Both China and Pakistan have

R. J. Augustus

increased their defense spending. Chinese expenditures have reportedly grown by 12 to 15 percent annually but is so negligible when compared to actual global expenditure (O'Hanlon ,1995). While Table 1 traces India's military expenditure from 1961 to 1996, Figure 3 represents the Indian defense expenditure for the year 1995-1996.

On the issue of Pakistan's and India's military spending, Pakistan's stability, integrity and strength are essential to India's national security interests. India's defense expenditure is not large

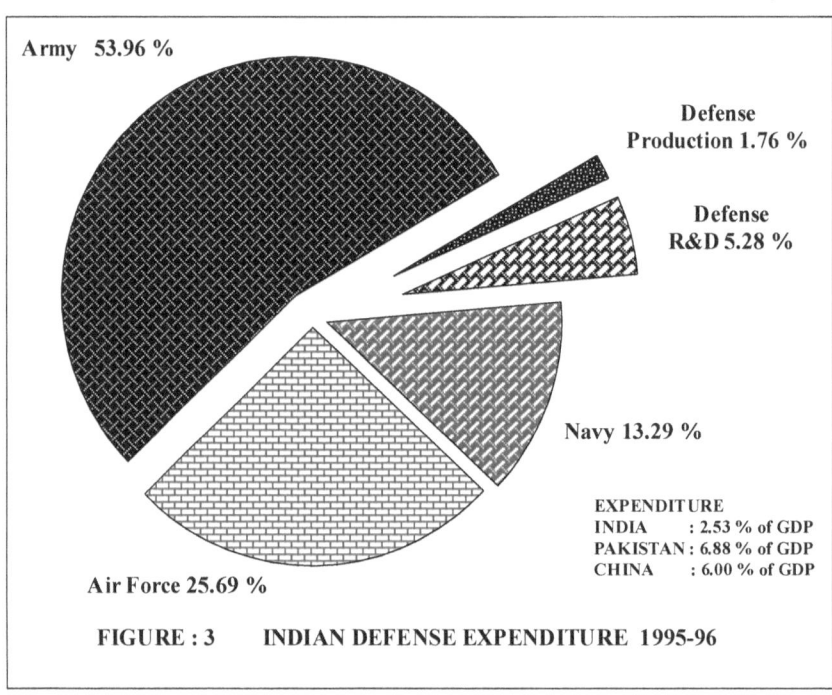

FIGURE : 3   INDIAN DEFENSE EXPENDITURE 1995-96

by global standards, but large enough to warrant a close scrutiny as to how it can be used effectively to catalyze development in other sectors. However, no Indian defense budget has ever been defiant. A look at Indian defense statistics for the past ten years detailing the annual defense budgets will confirm that the annual budgets have steadily declined.

Year	Rs. (billions of Rs)	$	Change %	Comments
1961-2	0.03	26		Defense R & Development Organization created (1958)
1966-7	0.12	73	+180	Wars with China (1962) and Pakistan (1965)
1971-2	0.21	99	+23	Indira Gandhi elected (I 967); War with Pakistan (I 97 1)
1976-7	0.49	130	+32	Arjun Tank Begun (1974)-, nuclear test (1974)
1981-2	0.96	170	+31	Indira Gandhi loses (1977); Regains office (1980)
1984-5	2.12	290	+20	Indira Gandhi Assassinated
1987-8	5.49	490	-5.8	
1991-2	6.83	290	-25	Economic reform begun under Narasimha Rao
1994-5	[11.75]	330	+6.8	Third Agni test succeeds after postponement
1995-6	[13.48]	[360]	[+9.1]	Plan 2005 approved

TABLE : 1 Indian Expenditure – Military R&D
Source: Government of India, Ministry of Defense, Annual Reports. Figures in brackets are approx. Others are Outlays - India Abroad, July 26, 1996

The 1996 defense budget again left little room for India's defense to grow (Raghuvanshi, 1996-B). This funding level has prohibited new procurement programs. Under the new budget the India Army has been allotted 56 percent of the total which amounts to $4.48 billion. The Air Force receives 26 % which is $2.1 billion, the Navy receives 14 % which is $1.11 billion and the DRDO receives 5% or $406.17 million. The Indian defense and Central government expenditure till 1997 is given in Table 2 and clearly indicates the defense expenditure as a proportion of the GDP has reached a plateau.

The Stockholm International Peace Research Institute, in its 'Arms Transfers database indicates certain trends in the import of major conventional weapons. In Figure 4 one can visualize the military expenditures (ME) of India, China and Pakistan as a function of their respective Gross National Product (GNP). There has been a slight tendency for India to overshoot China from '93 to '95 and has then come down. Pakistan's ME/GNP ratio is of course way above the other two countries. It is obvious that Chinese imports are way down because of their focus on their own production of conventional weapons. The trend indicator values have been telescoped together for comparison and is indicated in Figure 5. A quick understanding of India's defense trends can be obtained by studying Figure 6.

Act : Actuals RE : Revised Estimates BE : Budget Estimates	7th Plan 1985–90	1990-91 Act	1991-92 Act	1992-93 Act	1993-94 Act	1994-95 Act	1995-96 RE	1996-97 BE	8th Plan 1992-97
(As per cent of GDP)									
1. Fiscal Deficit	8.2	8.3	5.9	5.7	7.4	6.1	5.8	5.0	6.0
2. Revenue Deficit	2.6	3.5	2.6	2.6	4.0	3.3	3.0	2.5	3.1
3. Budgetary Deficit	2.1	2.1	1.1	1.7	1.4	0.1	0.7	0.5	0.9
4. Primary Deficit	4.8	4.3	1.6	1.3	2.9	1.4	1.1	0.2	1.4
5. Gross Tax Revenue	11.2	10.8	10.9	10.6	9.4	9.7	10.0	10.6	10.0
(a) Direct Taxes	2.1	2.1	2.5	2.6	2.5	2.8	3.0	3.1	2.8
of which									
(i) Income Tax	1.0	1.0	1.1	1.1	1.1	1.3	1.4	1.4	1.3
(ii) Corporation Tax	1.1	1.0	1.3	1.3	1.2	1.4	1.5	1.6	1.4
(b) Indirect Taxes	9.1	8.7	8.5	8.0	6.8	6.9	7.1	7.4	7.2
Of which									
(iii) Union Excise Duties	4.9	4.6	4.6	4.4	3.9	3.9	3.7	3.7	3.9
(iv) Customs	3.9	3.9	3.6	3.4	2.7	2.8	3.2	3.5	3.1
6. Total Expenditure of which	20.5	19.7	18.1	17.4	17.5	16.9	16.7	16.3	16.9
(i) Interest Payments	3.4	4.0	4.3	4.4	4.5	4.6	4.7	4.8	4.6
(ii) Major Subsidies	1.7	1.8	1.6	1.3	1.3	1.2	1.1	1.2	1.2
(iii) Defence	**3.3**	**2.9**	**2.7**	**2.5**	**2.7**	**2.4**	**2.4**	**2.2**	**2.5**
(iv) Other Non-Plan Expenditure	5.0	5.7	4.5	4.0	3.6	3.6	3.9	3.8	3.8
(v) Budget Support For Plan	7.1	5.3	5.0	5.2	5.4	5.0	4.4	4.4	4.9

TABLE : 2   Indian Tax Revenue and Expenditure
Source : Economic Survey of India, 1997

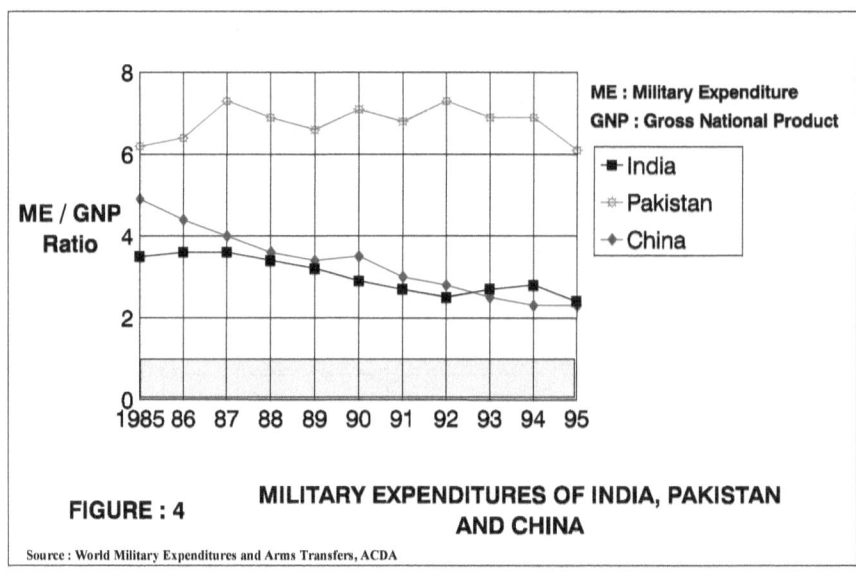

FIGURE : 4   MILITARY EXPENDITURES OF INDIA, PAKISTAN AND CHINA
Source : World Military Expenditures and Arms Transfers, ACDA

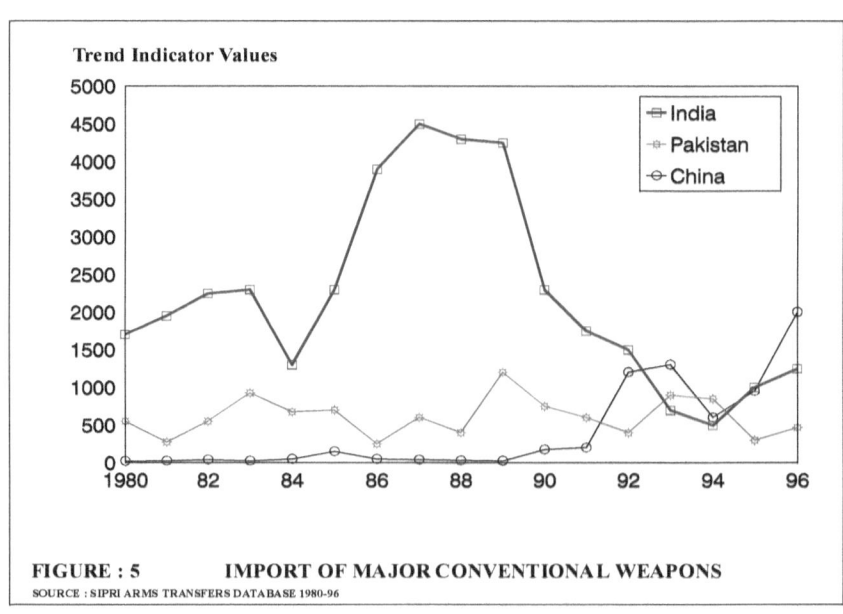

FIGURE : 5   IMPORT OF MAJOR CONVENTIONAL WEAPONS
SOURCE : SIPRI ARMS TRANSFERS DATABASE 1980-96

In Defense Of My Country

Emphasis has been placed on just the trend and not the exact percentages. Imports of different conventional weapons, the major suppliers of these weapons to India, the types of conventional weapons produced under license and the countries that have agreed to this licensed production are presented in this figure.

There is so much talk about India's defense endeavors for defending its own borders, but what should be of real concern is the significant rearming in the region, much of which is being done by the US, Russia and other major powers. To Indian criticism of the US rearming countries in the Middle East, the response has been that to the extent that the US is not able to prevent rearmament, it will pursue a balance of power policy to prevent any one state in the region from growing too powerful.

Sadowski feels that what is missing is not a vision of arms controls but a suitable means of putting it into effect (Sadowski,

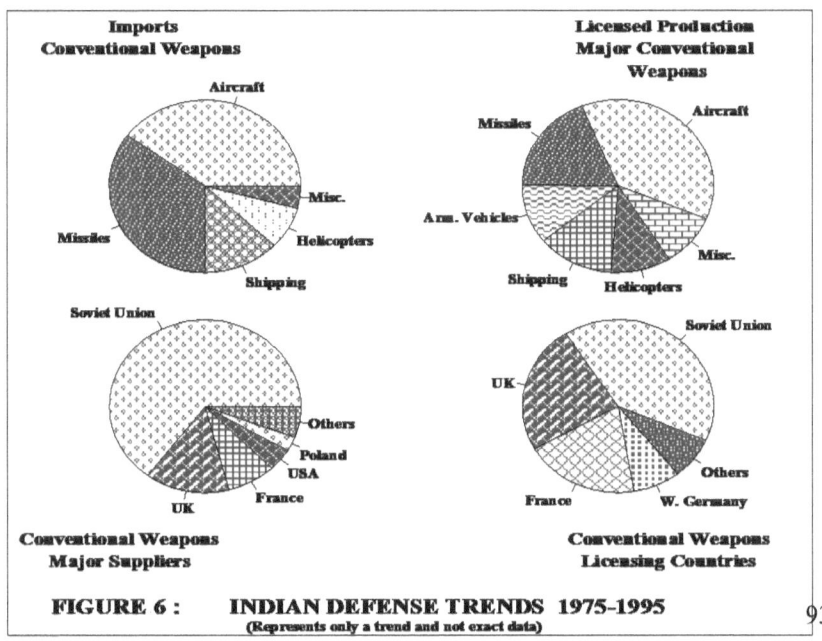

FIGURE 6 : INDIAN DEFENSE TRENDS 1975-1995
(Represents only a trend and not exact data)

1993). What is also disconcerting is that the US goes easy on its allies when talking of arms controls. A recent report by Dana Priest in the 14 April 1998 issue of the Washington Post describes how the US only addresses those countries they are worried about. Proliferation in countries like Israel, Saudi Arabia and Egypt rarely show up in any US government assessment (Priest, 1998). In the same article, Michael Moodie, president of the Chemical and Biological Arms Control Institute says that this neglect diminishes the authority the US might bring to the area of arms controls. The debate on selling supercomputers is another issue for the US (Milhollin,1998). Since the US seems to think so strongly of these issues, a look at the US strategy in the Asia Pacific region ought to bring into focus several misgivings. An understanding of the US strategy in the region should help India buttress her own strategic thinking.

6

The United States' Perspectives

To look at the scenario of Indo-US relations, it is necessary to study the fears and perspectives of both countries to come to a fair analysis of why each country behaves as it does on the international stage. Quite a bit of an effort has been put in by the US in tracking the continued proliferation of technologies associated with Weapons of Mass Destruction (WMD) and their delivery systems. US statistics reveal that approximately two dozen countries have ongoing programs to develop or acquire weapons of mass destruction with the US and other Western countries in the lead. Americans feel that as we enter into the next century many developing countries would have the capability to integrate nuclear warheads with indigenously produced long range missiles. At the same time the US also feels that at least for the next decade or two there may not be any country capable of or inclined to train and launch a missile to the continental USA. The US also understands that though it is doing its level best to slow the proliferation of these weapons, any country keen on acquiring such a capability will not have too much of a problem. But as far as conventional weapons systems development is concerned, the scene is not all that clear. Though defense establishments in major industrial countries do not have the kind of resources they once had at their disposal, they are still engaged in developing very advanced weapons systems but have a very difficult time designing, developing and fielding these weapon systems. This is also underlined by the fact that defense forces in most of the countries are being downsized or 'right sized'. In addition, since most countries are trying to get into the technology and weapons business and regional issues compound the risks involved, there are clear indications that the world is gradually becoming a very competitive and very unstable place and these tensions may soon spill over the regional borders.

So a good question to pose would be 'What does the United States have to fear from developing countries ?' If one may look at things from the US perspective, the first US concern is that developing countries are getting a foot in the high technology door. White (White, 1995) voices these concerns and states that developing countries are buying high technology for information warfare, drone and weapons of mass destruction. He says that these countries are looking towards psychological warfare and advanced business practices so that they can adapt and innovate in their own system . This seems to be a very familiar path followed by the US and her allies in enhancing their own technology supremacy. Yet developing countries feel that if they do the same, these activities become spooks that peep into Western windows. But these are concerns nevertheless and there is an unavoidable need by the US to keep a check on who is active, what the potential threat is going to be, and making sure that the US is not mesmerized by her own capabilities.

The primary concern of the US is the proliferation of nuclear, biological and advanced conventional weaponry. Vast quantities of advanced conventional weaponry and technical expertise are available for those who want and can afford it as a result of the collapse of the Soviet Union. This has not made things easier for the US and the threat perceptions have changed. Even though the threat of a full-scale nuclear attack on the United States has been greatly reduced there are other unknowns. The absence of a known threat creates a different type of fear – a fear of danger to US security where a deadly array of weapons and technologies in the hands of wrong nations and organizations may be used against the United States (Widnall, 1995-C). The American apprehension is that the regional problems in South Asia, Eastern Europe and former Soviet Union could pose a serious threat that would harm US interests, directly and indirectly (Carter et. al, 1992). The fear of proliferation has an impact on US security perspectives in a way that cannot be directly addressed through superior force or readiness for there is an intertwining of US interests with the interests of other nations. Even when US interests are not at stake, the United States feels responsible for the world order.

The US is also apprehensive about maintaining world technology supremacy and realizes that as high technology proliferates, their leadership could be a temporary one. Countries bold enough to challenge the US may also take advantage of the fact that it is US policy never to undertake any military action without some form of international consensus. In today's competitive scene, it is not unthinkable that the US might someday find herself alone in the world. From the technological stand point, the US depends very much on supremacy in the utilization of information and the associated technology needed for conventional power projection. Software is fundamental to information and many nations with India being the foremost, boast of extensive trained manpower. Also, the core technologies that support decisive information technology functions are available off the commercial shelf and third world access to these items cannot be denied. In like manner, materials required to make weapons of mass destruction are also available. The information easily available in the public domain is again knowledge that cannot be denied or hidden. All this because of their destructiveness and the fear they engender, could be used by an aggressor as a means of offsetting US conventional military power (Carter et. al, 1992). Carter argues that as a matter of logic, these radical transformations in the nature of the threat compel radical revisions in US security policies. The new security problems require more constructive and more sophisticated forms of influence other than mere readiness and deterrence. So for American security, the trend of technology investment and defense preparation in other nations is more important than any immediate challenge.

    The US forces have been accustomed to facing Soviet weapons in regional conflicts for a long time and the US weapons were designed and developed to counter Soviet weapons while US forces were trained with the Soviets in mind. In the future, US forces might face new technology and weapons manufactured in other countries that will be highly sophisticated and largely unfamiliar and therefore the US will not have the same preparation to counter these threats. This could well be very true of today's allies who at some time may decide not to play ball with the Americans. With a recognition of this threat, the US will view proliferation much more seriously and exercise extreme caution in all arms and technology transactions. The increasing technical capability of developing nations, the utilization of commercial off the shelf (COTS) technology for building defense systems and the easy availability is another fear that drives US strategic thinking. These developments in turn place much of the defense technology activity important to nonproliferation squarely in the marketplace. Another one of the problems for the US is the concern of losing out its technology leadership to other countries. So a combination of this free access to US technology by other countries, the proliferation of the technology, the tenacity and determination by developing countries to match the US have all joined together to make the US ultra-sensitive to the technology and arms transfer process. So one can understand why the US has taken this so seriously and is working towards making export controls and international weapon regimes a must in technology trade.

US Strategy In The Asia-Pacific

A future world war truly risks the annihilation of humanity and the reduction of such a risk is fully endorsed as what must be the first priority in security planning of any country. It is a general understanding that the US strategy for preventing conflict in the Asia-Pacific region has been to maintain a strong military presence and firm alliances with countries in that region. The foundation of this strategy is to keep the US-Japan alliance alive and Japan as a base for keeping an eye on the region. Dr. William Perry feels that this strategy seems to have worked well for the Americans and has allowed them to defeat aggression, deter war and guarantee peace (Perry, 1995-A). Military and economic 'aid' have gone hand in hand with much of the former disguised as the latter and the vast bulk of American 'economic' aid to the Third World has gone to a few countries of special strategic political military importance such as South Korea, Taiwan, Indochina and Israel (Frank, 1981). However, as far as South Asia is concerned, it attracts US attention and interest only during periods of crisis (Tahir-Kheli,1997).

R. J. Augustus

It may be argued that the US military presence and security alliances in the Asia-Pacific region are spill overs of the Cold War strategy and that it would be better for the US to withdraw her forces from the region and let countries define their own strategy and defend themselves. Though the US seems to have thought this through, they seem to have concluded that the consequences could be dangerous for the regions and also perhaps for US global strategy. One reason could be that for years, the US has provided a secure environment within which certain Asia Pacific countries have been able to build their economies rather than their defense. It must be acknowledged, even if it is grudgingly, that the US military presence in the region has in a way provided regional stability thereby creating a win-win situation for those countries and the US.

Some have advocated the idea of the US reducing its presence in the region and opt for multilateral security plans. While these would be confidence building measures that might work, it would also pave the way for countries to test the security waters, thereby causing instability in the region. If attempted, this would have to be a very precise tightrope exercise. The other alternative is not at all attractive. If the Americans left the region to its own devices, all regional equations would viciously change. The difference in economies would create jealousies thereby rekindling old animosities and to avoid the inevitable conflicts defense build-ups would commence. So priorities will change with defense buildup right on top of the list. This amassing of military might would have an impact on the Indian subcontinent. India would then be forced to follow suit with a proportionate or larger build up depending again on how quickly India adapts her strategy to this change. Pakistan and China would soon get into the act.

This is the sort of a chain reaction that any sane country will try to avoid. So if the US intends to avoid such a situation, then it must maintain a presence in the region. This could perhaps continue till multilateral relations mature over time.

US Technology Strategy

There is no doubt that technological superiority has been a cornerstone of United States security and industry for more than fifty years. Now most analysts are of the opinion that though this technological cornerstone is not crumbling, it has eroded significantly over the last decade (Gibbons,1989). Paige states that the US would like to retain the edge on current and future battlefields (Paige, 1995) even though today, unlike the early days of the cold war, the development and application of new technologies occurs globally. Any number of foreign companies have made deep inroads into high-technology markets that had been more or less the exclusive territory of US industry. The reduction in defense spending leading to downsizing and the increasing availability of high technology in the open market has caused a severe tightening of the economic belt for the defense industry . Simultaneously, the Department of Defense reports that defense technology in countries like the Soviet Union, France, UK and Israel are catching up with the US technology level even if some of it is derived from the US. One cannot blame these countries for getting these technologies if one goes by the mid April 1998 statement of US Secretary of Defense William Cohen saying that the United States was ready to fund R&D for a third battery of Israel's Arrow anti-ballistic missile. This is just a case in point.

R. J. Augustus

In the 1994 Department of Defense report on the US Technology Strategy, John Deutch says that the technological superiority of US forces contributes to reduced casualties and quicker victories. Americans feel that this superiority must be maintained today under circumstances profoundly different from those of even the recent past. When the US faced the Soviet Union, they faced a technologically competent, largely predictable single adversary and evaluated technological progress and rate of modernization on a one to one basis. The end of the era has logically called for a reduction in defense resources. So the US must put aside all the old and expensive military specifications and turn to the commercial sector for all her defense needs - for it is here that one can find cutting edge technologies being honed by prudent business practices. This leaning towards the commercial sector is not the prerogative of the US alone since other countries too are following suit. This means that the US must also develop new approaches that will enable her to maintain the edge over other countries even if they have the same access to basic high technology.

These sensitive and conflicting issues are addressed by the US Department of Defense by involving the user of the technology early in the program and more often in technology development in order to reduce the time taken to field defense systems with always an eye on reducing cost and improving system performance. These and other innovations comprise the US Defense Science and Technology Strategy, and its accompanying Defense Technology Plan, which is the blueprint for executing the strategy. While this represents a coherent, well thought out plan for keeping the US national security technological edge in a time of profound change (DOD, 1994), one must also realize the extreme competitive nature of the defense technology game and that other western countries are almost up to speed in the game.

On the political, military and technological aspect, the US administration is extremely keen on extending its cooperative international efforts. In the case of defense weapons development and production they have worked towards cooperation with their allies to take advantage of joint development of expensive new systems. Apart from being attractive from the economical and operational aspects, it also ensures that the respective defense industries and establishments are woven together in a way that strengthens and is mutually supportive . The theater ballistic missile defense cooperation endeavor could be cited as one of the best examples for strengthening political and military alliances.

American attention today, has been drawn towards the economic health of her industry and a reassessment of the global security environment. This reassessment has led to defense cutbacks resulting in a dramatic effect on the US high-tech defense industry. This is especially true of the aerospace industry because of the government reluctance to fund major projects. This hesitation is paving the way for the government to relinquish its leadership role. So there is a great sea change taking place in the way the US industry and US government will form a partnership for the future.

The US defense programs were geared to produce sophisticated and superior weapons and this technological superiority was driven by the need to counter an enemy who could field weapons of almost equal performance. While threats have changed, the challenge, today, is to increase the soldier's access to new capabilities at a fraction of the cost of earlier systems in order to take into consideration the turnover of new commercial technology. Since commercial technology must equip the defense forces and support current needs and solve defense problems of new missions and threats, the US considers it 'imperative to avoid technological surprise, which historically derives from the integration of technologies into dramatic new strategy and tactics' (DOD,1994). Hence, the United States has been very sensitive in protecting its interests on these fronts, and extremely active in arms controls and raising embargoes on technology transfer to countries that it fears will at one time or another pose some form of a threat to herself.

In Defense Of My Country

7
Arms Control

Conventional Arms

Most countries have more or less agreed that unlike weapons of mass destruction, which should be strictly regulated and eventually eliminated, conventional weapons may be considered as a legitimate means with which countries are entitled, in time of need, to exercise their sovereign right to individual or collective self-defense (Donowaki, 1995). There has been a fair amount of interplay and relaxation between giving up or prohibiting weapons of mass destruction and the continuing availability or usage of conventional weapons. So one may say that conventional weapons have not been treated to a rigorous global control. According to Janne Nolan in the book "Cascade of Arms: Managing Conventional Weapons Proliferation', conventional weapons restraint as it is today cannot boast of even the most rudimentary diplomatic infrastructure. The decline in weapon imports because of own production in developing countries is testimony to the fact that conventional arms can be legitimately developed, produced, deployed and traded by any country which is a member of the United Nations.

It would not be practical to analyze the problems associated with practicing export controls without referring to the driving forces behind those who impose the export controls and those who rebuke them. Drace (Drace, 1994) states that a country's strategy, morality and economic self-interest play a notable role in the exercise of export controls. It is also true that the practice of export controls in any country depends on basically identifying the enemy, defining the issues involved in dual-use technology, considering differing national situations and ensuring the controls are effective and above all being even handed in all dealings. Above all it must be remembered by those who work the day to day issues of export controls that these controls should not be blindly applied each time an export license request is made. If that were the case and discretion was not meant to be used, the whole decision process would have been computerized a long time back.

With that preamble, one can examine the paths which the exporting countries may follow in-order to discourage proliferation of dangerous weapons and regional arms races without falling out with the countries involved. The first path to be followed is to encourage the peaceful resolution of regional conflicts. Countries in conflict must be asked to restrict the category and quantity of weapons systems being imported or produced, for, ever so often it is seen that when countries are determined to obtain weapons and weapon technology that they believe are imperative for their security, they will obtain them by any means at their disposal. The two straight options for them would then be to either purchase the equipment or produce it indigenously. It is when the decision is taken to procure the weapons and technology that exporting countries can apply the controls and then we get back to where we started. On the other hand, if they decide to produce the weapons, it would ultimately lead to themselves exporting the systems for economic reasons and risking the title of 'proliferator'. Global efforts may slow such proliferation, but rarely stop things completely. We are all too aware of countries that have turned to exporting such systems and anyway that is how arms exports started. In this regard India stands way above her peers. Even though several offers have been made to India to purchase its missile technology, India has flatly refused to sell its technology even though the sales would have been most welcome to support indigenous Research and Development.

International agreements are another means of helping countries help themselves. Countries like the United States that have international leverage could provide the support for these commitments and ensure that these commitments are adhered to. However mention must be made of the fact that India and the United States are among the few countries that basically adhere to any agreement, while some countries will sign any agreement knowing that they have no intention whatsoever of adhering to the guidelines.

There could be several reasons why an arms exporting country would like to restrict its exports. Since we are specifically dealing with the United States one would like to understand the US reasons for applying such restrictions to, say a developing country. Very simply put, there are two basic reasons. Firstly, in some cases the U. S. may be the only country possessing the technology or weapons system, and it is hoped that withholding that capability from a developing country slows its technological growth and provides the US with a technological edge over potential adversaries. Secondly, a particular country may be so incorrigible or dangerous that the US wishes to distance itself from that country. Sometimes, this restraint by the US has not been followed by its allies and lucrative markets have simply been taken away from US suppliers and promptly been filled by the allies.

R. J. Augustus

Developing countries not able to receive technology from the US are today looking for second sourcing in Europe. Exporters have argued that if a commodity is available from foreign sources that do not have comparable export controls, US export controls serve no purpose, since objectionable users can obtain the items elsewhere and continue unhindered with their weapon programs. Those stressing the benefits of controls, then, argue that some economic sacrifices (in the form of reduced exports) are worth the price. They say that if exporters are burdened by controls, the burdens should be seen as part of the price of doing business with potentially dangerous commodities. But this is where the US administration must take a practical view of its export control policy considering her own industry's demands and the approach by other western nations. Kopte (Kopte, 1996) explains that one will witness in the coming years, even more second-hand weapons and equipment being offered for sale. This will be true in the United States and Russia where surplus stocks include a large number of highly sophisticated weapons are even now being given away for free or at very low prices. In practical terms, this reduces the revenue that can be earned, but it also avoids or reduces the cost of mothballing or scrapping the weapons .

The international stage was for so long driven by the Cold War objectives that today it requires a great effort to get out of the rut and change basic philosophies. The application of new formulae for export controls in the changed security environment is one of them. Regional irritations have mushroomed and there is a gnawing fear that these disputes are already linked to the proliferation of weapons of mass destruction, leading to instability not only in these regions but in the whole world. With the advent of such threats, Watanabe (Watanabe, 1994) confirms that the prevention of the proliferation of weapons of mass destruction, their delivery platforms, and the prevention of the excessive build-up of conventional arms have become major global security concerns and consequently, the importance of export controls is now internationally recognized

While non-proliferation export controls are aimed at preventing the proliferation of weapons of mass destruction, missiles and the excessive build-up of conventional arms, in day to day terms and the way it is being applied, it could cover pins to pianos. If such mindless export controls obstruct economic and technical development of countries subject to control, this will invite their economic and technical isolation, which could actually be a negative factor for regional security. In addition, if there is a restriction of technology transfer under the assumption that the dual use technology may be used for developing weapons of mass destruction, one is starting out with the premise that the nation is guilty until proven innocent – contrary to fundamental US philosophy. If such is the approach to legitimate technology requirements, developing countries will always stay as developing countries or at best their progress, if any, will be very slow.

Therefore export controls must be tailored for different conditions of country and legitimacy of requirement. A view of some of the control regimes will enhance the argument that international technology transfer has not been approached with much prudence. At best they have a parochial ambiance and not a global one. Therefore, non-proliferation export controls should be implemented with restraint so that they will achieve their objective, and should not retard the economic and technology development of countries subject to control.

R. J. Augustus

Non-Proliferation Regimes

In general, proliferation would specifically refer to the trading in 'dangerous' nuclear, chemical and biological warfare and missile technology. There are a growing number of people who argue that the definition is too restrictive as there are a variety of other high-technology capabilities like regional surveillance, advanced communications, stealth, precision navigation and smart targeting which together have the potential for being major force multipliers. For example, modern high-performance aircraft may be as effective as missiles or more so in delivering nuclear weapons. So a good question for India to pose to the United States would be to ask why should there be the single-minded focus on missiles when the two countries have a fairly good cooperation in aircraft technology.

In recent years, there has been a better synchronization in efforts of ad-hoc export control regimes in dealing with dual use technologies, goods and equipment. The primary reason for this development is the projection of proliferation as the major threat today to international peace and security. A common factor in these regimes is that these are not treaty obligations but understandings that are then implemented by individual countries. This is demonstrated by the fact that only the US imposed sanctions on the Indian Space Research Organization (ISRO) and the Russian Glavkosmos because of the Cryogenic engine deal that was at cross purposes with the MTCR obligations.

Missile Technology Control Regime

The modern missile is one of the dominant military technologies of our time and has acquired great strategic and political significance. Armed with conventional high explosives, they are a cheap and effective way of taking out enemy targets and armed with nuclear or chemical warheads, they become weapons of mass destruction. They have a lot in common with space exploration, but the mandate of the United Nation's Outer Space Treaty (OST), and the G-7 nations' Missile Technology Control Regime (MTCR) rarely share the same platform. While the 1967 OST bestows on all countries the right to peacefully explore and use outer space, the 1987 MTCR instituted international coordination of expert controls on weapons delivery platforms, with the goal of controlling nuclear proliferation. Since the technology used in space exploration can be definitely used in a missile program, thereby defining itself as dual use, in the context of export controls a definite conflict arises (Hurewitz, 1994).

     Deborah Ozga, in 'A Chronology of the Missile Technology Control Regime (MTCR)' (Ozga, 1994) explains how the MTCR announced its guidelines on 16 April 1987, after years of meetings by seven western countries. According to the guidelines, the MTCR's original purpose was to reduce the risks of nuclear proliferation by placing controls on equipment and technology transfers which contribute to the development of unmanned, nuclear-weapon delivery systems. Over time, that goal was expanded to limit the risks of proliferation of weapons of mass destruction by controlling transfers that could make a contribution to delivery systems for such weapons. Some countries feel that the formation of the MTCR was a clandestine affair that came into being after four years of secret negotiations among the industrialized countries. At that time India had not deployed any ballistic missile, even though it had a close defense relationship with the Soviet Union. Joshi (Joshi, 1994) however states that not unlike the Pokharan test of 1974 that sparked off the London Nuclear Suppliers Group, the SLV program was one of the factors that made Western governments aware of the India's missile capability.

In some circles there is a deep feeling that India may have played a larger role in the evolution of US views on missile proliferation. This is because of India's entry into their exclusive space club by placing a tiny civilian satellite, Rohini I in orbit and the Indian government's formal sanction of the Integrated Guided Missile Development Program. Depending on how one views the program whether in the context of the DRDO or the Indian neighborhood, it can be considered to be technology or threat driven. The best indicator of this according to Karp (Karp,1991) has been the fact that though India acquired surface-to-air missiles and multi-barrel rockets from the Soviet Union, it never went in for longer range systems. India's missile progress, its alleged nuclear capabilities and refusal to be pushed by the US into signing arms control measures are the some of the main irritants in the relationship.

While India thinks that the MTCR was aimed at her missile program, China thinks that it was aimed at her capabilities. Hua (Hua, 1991) says that China exhibited the M-family of missiles and openly said that they wanted to sell it. It was a public exhibition and then suddenly, the next year, the MTCR was started. China's export of ballistic missiles was transparent and it was the US stick that forced China to go underground.

MTCR - Strengths And Weaknesses

The international community considers the spread of missiles and missile technology as a crucial security issue and has been singled out in the United States for more stringent export control than is applied to other technologies. The MTCR provides the international norm, for dealing with this proliferation and while it did have its strengths for some time, there are inherent weaknesses and its capability is now being questioned. It is only the advanced nations that have imposed such controls on the smaller countries and Janne Nolan explains that the main purpose of this approach is that industrial nations can still exert decisive influence over the defense programs of developing countries by imposing controls on technology transfer (Nolan, 1995). Developing countries have chafed under these restriction and complain that the regime is inherently discriminatory while being unverifiable and unenforceable (Kemp, 1992). While the MTCR members are unruffled by such criticism, one must remember that many developing countries were able to and are still developing their own systems with help from the US and European countries.

The principal faults of the MTCR is its definition of nuclear-capable missiles, the unrealistic guidelines for range and payload, the possibility that suppliers may sell missiles that fall within the MTCR's scope with the assurance they will not carry nuclear weapons. To make matters worse, the strict application of the MTCR mandates by the US has unnecessarily restricted the export of required dual-use technologies and is "throwing the good projects out with the bad" (Hurewitz, 1994). The MTCR has no international agency to monitor compliance, no enforcement mechanisms, and no institutionalized arrangements for regular meetings among participants. So even with its modest scope, the MTCR has been mired in some serious disputes and controversies over its implementation virtually since its inception (AER, 1991). The controversy over France's export of liquid fuel motor technology to Brazil in the late 1980s and Russia's deal with India on the Cryogenic Engine in 1992 are arrows in the quiver for the opponents. The most important issue and the fundamental weakness of the MTCR is the small number of adherents.

Whether or not India decides to develop or deploy ballistic missiles has now gone into the maelstrom of politics. The DRDO provides India with the option of fielding short and long-range missiles . American arguments that such systems will be destabilizing cannot be taken too seriously since, in their simplest form, they seem to argue that missiles in their own hands promote peace and those in the hands of the others do the opposite. However, American pressure and the ability to influence Indian policy cannot be underestimated. They have been able to delay the Agni program and have created a kind of defensiveness in the minds of Indian policy-makers over the Prithvi that the government is unable to effectively counter (Joshi, 1994).

Non Proliferation Treaty

The Nuclear Non-Proliferation Treaty (NPT) is the foundation of the nuclear non-proliferation regime and has served as the principal legal barrier to nuclear weapons proliferation for more than twenty years and with such a vast number of member states, that it is the most sweeping, comprehensive, and by western views, probably the most successful technology control regime in existence. However the NPT has divided the world into the nuclear 'haves' and 'have-nots' with the purpose of halting the spread of nuclear weapons technology in exchange for promoting the spread of peaceful nuclear energy technology. The dual standard of nuclear ownership of nuclear "haves" and "have-nots" has been recognized by many states as inherently discriminatory, and contributed to the impression that US nonproliferation policies were unjust (Martel, 1995)

The NPT has not been successful, even among its own member countries, in completely stopping the spread of nuclear weapons, in spite of its success in declaring the non-nuclear status of most of the countries of the world. Most western nations believe that Israel, India, Pakistan, Brazil, and Argentina all remain non-members with significant nuclear programs, though Brazil and Argentina signed a similar agreement with the IAEA in December, 1991. Iraq is a party to the NPT but was nevertheless engaged in a sophisticated clandestine program to develop nuclear weapons, causing deep concern over the effectiveness of IAEA inspection procedures. The International community feels that it cannot afford to be complacent because the threat of proliferation of weapons of mass destruction and their means of delivery has received a new impetus in spite of the end of the strategic nuclear arms race between the superpowers (Kurosawa, 1994). In recent years, France, China and South Africa among others, have acceded to the treaty, making it more universal.

The breakup of the Soviet Union has had a double impact on US-Indian security relations. Firstly, it has diluted the close defense relationship between New Delhi and Moscow and secondly both India and the US share concerns about the availability of nuclear weapons in the new republics of Central

Asia. India is doubly concerned since now she is almost surrounded by nuclear powers. For these reasons, it is not surprising that Indian officials have begun to reassess their policy on proliferation and defense cooperation with the United States (Kemp, 1992). Surprisingly for India, even Russia wants her to sign the Nuclear Non-proliferation Treaty (NPT) and the Comprehensive Test Ban Treaty (CTBT), both regarded by New Delhi as discriminatory (Reuter, 1996). The traditional position, with many advocates in the broader community of non-proliferation specialists, is to insist that both India and Pakistan sign the NPT as non-nuclear weapons states and eliminate their nuclear weapons capabilities accordingly. This policy, while reflecting an optimal objective, is deeply flawed at the present time (Perkovich, 1995). The NPT policy fails to recognize how thoroughly opposed Indian officials and citizens are to signing a document that divides the world into two classes of power. India like Israel is surrounded by not so friendly countries. Yet the US has been more accommodating to Israel on this point, which tends to further deteriorate Indo-US diplomacy. From where India stands, she is forced to realize that a nuclear capability would place her at par with the world powers and it is difficult to ignore this symbol of sovereignty especially during the elections. India had never endorsed nuclear weapons as weapons of war and had steadfastly maintained that nuclear weapons must be totally banned and eliminated. Proliferation has always been India's concern and was equally opposed to proliferation by new nations. It set an example by not building up an arsenal after its nuclear test of 1974. Unfortunately, as Subrahmanyam says, 'a world conditioned by nuclear fundamentalism could not understand the Indian perspective and misinterpreted it' (Subrahmanyam, 1992).

In Defense Of My Country

The treaty has a major flaw as the accusation of discrimination is not addressed by it. While the third world has criticized this, India has been particularly vociferous in accusing the nuclear powers of trying to prevent the developing world from gaining access to sophisticated technology relevant to its development. The NPT has also been criticized for being partial towards nuclear weapons acquisition in some states, such as Israel. Indians feel that Americans often act as if this discrimination should be accepted as a fact of life. But for Indians especially, discrimination is precisely the fact of life that triggers much of their interest in nuclear capability. Saying no to the treaty represents equality with the nuclear powers, independence and a feeling of national power.

However even if there have been opposite views of looking at the NPT (Dunn, 1990), one may under these circumstances be excused for concluding that this international regime for controlling the export of nuclear weapons and the accompanying technologies, is no longer sufficient to prevent all nuclear proliferation (ECO,1991; Kapur,1990). The evidence states that the conventional tools for controlling access to nuclear technologies and weapons is extremely limited and cannot prevent countries from developing and demonstrating their nuclear capability.

Comprehensive Test Ban Treaty

Though the Comprehensive Test Ban Treaty (CTBT) is an issue that India proposed to the world more than forty years ago, there was no channeled approach towards it till recently. It was first proposed as an arms control measure in the mid-1950s when US and Soviet atmospheric nuclear tests proclaimed to the world the gross hazards that the massive destructive power of nuclear weapons carried with it. The CTBT has been the goal of the international community ever since these tests were banned in 1963. The 1970 NPT called for a comprehensive test ban treaty and since then the introduction and signing of a CTBT has been given paramount importance. Serious talks on a CTBT began only in 1994 at the Geneva Conference on Disarmament (CD), after the United States, Russia, Britain, France and China decided to form a negotiating committee.

As in other countries, even in the US there are those who support and those who oppose the US signing any form of a test ban treaty (Arnett, 1988). Article VI of the Non-Proliferation Treaty (NPT) states that countries already possessing nuclear weapons are obliged to take measures to cease nuclear arms buildup at the earliest. Several non-nuclear weapons states have given up the option to produce nuclear weapons by signing the treaty because of this clause. Mohammed Shaker, Egypt's UN ambassador and president of the 1985 NPT review conference emphasizes that, "A comprehensive test ban is essential for the viability of the NPT."

Though the nuclear test ban has been an arms control objective for more than 40 years, it was felt that only after years of negotiations was an agreement on a Comprehensive Test Ban Treaty (CTBT) within reach (Kimball, 1996). Time ran out for the negotiations and new obstacles developed. Although the aim of nuclear weapons elimination has been officially endorsed in various ways by the governments of all the world's major nations, including the five declared nuclear powers, there was little agreement among them on the forum, timing and conditions for advancing disarmament. France had completed all the tests that it wanted to and China conducted tests even in 1996 while her adherence to the treaty was considered vital to securing US Senate ratification of the CTBT.

The Indian Government had indicated at the highest levels that while it remained committed to the objectives of the CTBT it would like the treaty to be an integral part of a decisive global movement towards the elimination of nuclear weapons. More than 60 percent of urban Indians surveyed in December 1995 by India Today magazine said they would approve if the nation conducted another test blast and 72 percent rated "protecting ourselves against nuclear threats from China and Pakistan" as the most important reason to have a nuclear program. The public options and nuclear options poll of the Kroc Institute Survey Opinion (Smith J.G., 1996), has shown that public opinion is very much in line with that of the Indian government decisions. Western countries have failed to allay India's fears and have not met Indian requests for a modification of the treaty. No one believed that India would turn its back on the CTBT as it did since it had the blessing of all the governments till then. It was felt that if India stayed away, it would be a terrible blow to India's prestige on the international arena according to Michael Krepon (Raja Mohan, 1995). The Indian government's official stance is that it would not build a nuclear weapon unless threatened. On March 20,1998 the new Indian government vowed to be transparent while spelling out the thrust of its policies and the defense minister George Fernandes after assuming office clarified that India would keep its nuclear options open but this did not necessarily mean that New Delhi would build atomic weapons. But one must accept that over the years, the reluctance of the big powers to disarm and a perceived threat from Pakistan and China have put pressure on India to join the club (Graves, 1996). To add to this, Pincus (Pincus,1998) writes that 'while the Clinton administration urges the Senate to ratify treaties that end nuclear testing and sharply cut the number of US and Russian strategic nuclear weapons, US government scientists are pressing ahead with new methods for keeping thousands of strategic missile warheads and bombs reliable and accurate for at least 25 more years'.

The Clinton administration had been counting on India's support (Krishnaswami,1996) for the CTBT but the progress was "painfully slow" (Naughton,1996). India feels that the NWS do not want to give up their nuclear hegemony and though the CTBT draft text was modified to accommodate China's concerns, India's concerns were not considered. The explanation given by western representatives is that the modification was made because China is a declared nuclear power. Though some consider that a bit of progress was made, the treaty is in a state of limbo (Nebehay,1996). India has not accepted the 'constraints' on its nuclear option since nuclear weapon states continue to rely on their nuclear arsenals for their security even though India's commitment to disarmament remains the same (Basu, 1996). In New Delhi, however, there is growing recognition that these so-called 'non-discriminatory measures' are aimed at capping Indian capabilities through the global route and justifications for and against this stand have been made (Bidwai,1996; Subrahmanyam,1996). In return, US Secretary of State Warren Christopher is said to have told the Congress that he had warned India that it would be "isolated" from the world community for blocking the CTBT (Haniffa,1996-C). This attitude is bound to echo in the Indo-US defense cooperation and in India's concerns for technology transfer for some time to come.

A world of nuclear double standards cannot be the basis on which a sane and secure world order can be erected. There is also a growing number of credible scholars who have come around to the realization that the NPT in its present shape will not stop proliferation, and are articulating these views at public forums, even though such views go against their governments' position (Jasjit Singh, 1995). To some, the CTBT appears pointless from certain aspects because through more sophisticated technology, the nuclear weapon states would be able to continue to test and develop new generations of nuclear weapons. What is needed as Apotekar clarifies (Apotekar, 1995) is a true comprehensive test ban treaty prohibiting all types of nuclear explosions in all environments such as underwater, underground, in the atmosphere, and in space. It should also ban all tests, for all time, including computer modeling and simulation.

Wassenaar Arrangement

The Coordinating Committee for Multilateral Export Controls (COCOM) was the first export control regime set up in 1949 by NATO countries with the objective of blocking exports of military equipment, munitions and dual use technologies to USSR, China and their allies. With the end of the Cold War, COCOM has now been replaced by the Wassenaar Arrangement. The Wassenaar Arrangement on Export controls for Conventional Arms and Dual Use Goods and Technologies was launched in The Hague on 19 December 1995. The Arrangement was formally established by a Final Declaration issued by the representatives of 28 participating governments.

R. J. Augustus

The basic objective of the Wassenaar Arrangement is to prevent the acquisition of conventional arms and sensitive dual-use technologies for military end-use by countries whose behavior is, or becomes, a cause for serious international concern. It is designed to complement existing weapon control and non-proliferation regimes and is not intended to impede bonafide civil transactions. The agreement provides for information sharing on both licenses issued and licenses denied. Admission of new participants will be decided by consensus and factors for consideration will include a country's adherence to NPT, BWC (Biological Weapons Convention), CWC (Chemical Weapons Convention), Australia Group, MTCR and Nuclear Suppliers Group.

The Wassenaar Arrangement is a work in progress and member nations are looking at ways to increase the level of visibility into arms sales and control mechanisms. There will have to be a lot of work carried out on the Arrangement before it is honed to be a better arms exports regime for it will have to satisfy its critics who consider it a poor substitute for the Cold War-era COCOM. The working group is also going to explore ways to broaden the membership of the Arrangement. At the moment the Arrangement cannot be used to stop specific arms deals, but is designed to tackle transfer trends rather than individual transfers.

The 20 January 1997 Defense News carried a commentary on the weakness of the Wassenaar arrangement by suggesting that the Arrangement should have provided a mechanism for regulating arms transfers that might upset regional stability. France, which has its own ax to grind in arms sales was one of the most staunch opponents to giving Wassenaar the clout to restrict such sales. As a result, the Wassenaar Arrangement, conceived as a forum for instituting arms export restraint, merely provides a forum for exchanging information on arms shipments. As a result of such decisions, the effect of weakness is clear and the US might have to battle alone.

The Groups

The other regimes of importance to Indian concerns are the Zangger Group, the Nuclear Suppliers Group and the Australia Group. The Zangger Group and the Nuclear Suppliers Group (NSG) are two important international nonproliferation organizations for nuclear nonproliferation. The Zangger Group was founded in 1971 by a group of major nuclear suppliers basically to understand the treaty and the precautions on exports of nuclear equipment and materials. In 1974 the Zangger Committee published a "Trigger List", that is, items which would "trigger" a requirement for safeguards. This 'Trigger list' known as the Zangger List was drawn up of specific nuclear equipment and materials and would not be exported unless the importing country arranged international safeguards for them.

When India tested a nuclear device in 1974, the nuclear weapon states felt that India was misusing nuclear technology and promptly the NSG was created. This event brought together the major suppliers of nuclear material, non-nuclear material for reactors, equipment and technology who were members of the Zangger Committee as well as countries who were not parties to the NPT. Also called as the London Club, the NSG adopted the entire Zangger List and added heavy water and heavy water production plants to it but for some reasons it was not active from 1978 and 1991 even though its guidelines were in place. In 1992, after the discovery of Iraq's secret import of nuclear technology with a lot of foreign help (Albright,1992), NSG expanded the list to include 65 dual-use items. Some countries have questioned the real motives of the Nuclear Suppliers Group. They are of firm belief that the clandestine manner in which it was formed was specifically meant to keep out developing countries. There are also those who believe that move was a clear attempt to curtail transfer of nuclear technology to developing nations.

To most western nations, the most difficult defense and foreign policy challenge today is the prevention of proliferation of WMD, chemical weapons in particular because of the ease in obtaining required materials in the commercial market. An attempt has been made in the form of the Australia Group which is a collection of countries that began working together in the mid-1980s to suppress the proliferation of chemical and biological weapons through export controls. Given the threats that this proliferation carries, it was generally felt by the members of the group that the mechanisms to control the flow of dual-use materials and equipment to proliferating countries receive support in all quarters. However the reception has not been very forthcoming due to the confusion between the operating features of the Australia Group and the Chemical Weapons Convention (CWC) even though the CWC and the Australia Group complement each other. For those detractors who feel that these restriction will effect trade and there is nothing to gain from this group, the explanation given is that since the Australia Group was started, statistics show that the overall chemical trade has actually increased. In addition non-members must take note that since 1918, all incidents of battlefield chemical weapons use have taken place in developing countries. So with the fear of such threats, it should be evident that no signal export mechanism is sufficient to address the problem of chemical weapons proliferation. To successfully combat such proliferation, the CWC and the Australia Group must have the support of suitable defense programs, program of education for the masses on such threats, strong intelligence and above all support by all countries that take these threats seriously.

One cannot cast blanket aspersions on the South Asian region for they too have made attempts to lessen the historic tensions. The South Asian Association for Regional Cooperation (SAARC) was established in 1985 with seven states, Bangladesh, Bhutan, India, Maldives, Nepal, Pakistan, and Sri Lanka as members. Its purpose has been to promote the welfare of the peoples of South Asia, to strengthen collective self-reliance, to promote active collaboration and mutual assistance in various fields, and to cooperate with international and regional organizations. The Governments of India and Pakistan too have tried to breach the gap. The India-Pakistan non-attack agreement signed in 1988 and entered into force in January 1991 and again, the chemical weapons agreement that India and Pakistan signed in 1992 have helped ease some tensions in accordance with international expectations.

India's Approach To The Regimes

The nonproliferation policies of Western powers are founded on the strategy of preventing and controlling the Third World development of technologies that could impinge on their military and economic interests. In helping create an expanding web of Western technology export controls, the United States has worked with the very industrial powers with whom it has been engaged in a fierce technology competition. Since all advanced technologies have dual applications in the military and civilian spheres, the present export-control strategies carry the danger of severely curtailing the flow of technology at the cutting edge of civilian modernization and development.

The regimes have definitely had an impact on India and her approach towards research, development and strategy planning. The MTCR has had a visible impact on India's civilian space and ballistic missile programs. Export controls and barriers to Technology Transfer have forced Indian planners to revise their project timetables and a number of programs have been delayed since a host of items and technologies have been denied to India under these controls.

The MTCR, former COCOM controls and Western national-export barriers have compelled Indian space and missile planners to locally develop the components and equipment denied to them. These efforts have yielded both success and failure. In general, the forced indigenization of items and technologies has slowed down projects and soared budgets. In India, the additional indigenization activity because of these controls is proving to be costly. However, the controls are also promoting Indian self-reliance in an array of strategic allied technologies which is bound to pay-off in the long run.

As quoted by R.P. Singh (Singh R.P., 1990), Geoffrey Kemp, Senior Fellow at the Carnegie Endowment for International Peace says that, "In some ways the situation is analogous to the efforts by the advanced powers in the 19th Century to deny tribes of Africa and the peoples of the Pacific Islands firearms and liquor on the grounds that they would mishandle them." What they are trying to stop is the emergence of new nuclear players. S.K. Singh says that the West is convinced that the Third World powers are not fit to handle nuclear bombs with any degree of responsibility and this really means enabling some countries to possess nuclear weaponry, while ensuring that the rest of the world remains without these options (Singh S.K., 1995). Japan and Germany too have become rather silent on certain aspects of this issue, ever since they have seen the possibility of becoming Permanent Members of the Security Council. Yet even the US has been helping selected countries to develop indigenous defense industries and has not stopped testing its missiles. One of the most expensive 30 minutes in weapons testing will occur in the fall of 1998 (Graham,1998). Arms aid, sales and military technology transfers or co-production defense contracts have been used as foreign policy instruments in states like Israel, Taiwan, South Korea and, recently, the People's Republic of China. Apparently, military technology export controls serve as a useful currency for procuring influence in the Third World.

The extent of Indian space or missile technology sales to countries suspected of proliferation will be determined, in part, by perceptions of international control regimes, India's export control system and more generally, Indian views of the proliferation problem. For the purpose of this book, it is especially important to note Indian attitudes toward the MTCR.

India is not a member of the MTCR nor has it given any specific public indication that it would readily identify with the regime's guidelines and parameters. India's objections to the MTCR closely parallel its often repeated objections to the NPT and other international supplier cartels. Opposition to what one analyst has termed a "global hegemonistic institution", centers around several inter-related issues. Indians see the MTCR as a "discriminatory" regime that allows some nations to have a certain class of weapon system that others are forbidden to have. The MTCR ignores regional factors and legitimate security interests of non-member states. New Delhi does support "non-discriminatory" regimes that involve weapon system prohibition for all states. India is a party to the Biological Weapons Convention and has expressed support for a similar ban on chemical weapon stockpiling and technology transfer precisely because the regulations in such regimes prohibit all states from possessing such weapons.

India also believes that the MTCR takes too narrow an approach to control the global arms race as it seeks merely to control the flow of weapons-related technology and does not consider the basis of interstate conflict. It also does not seek a global disarmament approach as a broader and thus more effective solution to the arms control problem. In addition, the MTCR seeks to control the transfer of certain dual-use technologies and thereby impedes the ability of Third World countries to access such technologies for non-military programs and hinders their economic progress. Most developing countries also object to the "self-righteous" attitudes of developed countries that attempt to apply arbitrary norms embodied in the MTCR on states who were not part of the cartel that established those norms at the outset of the accord. Indians are quick to note that it was not they but the Western countries who transferred dangerous technologies to Iraq. In other words, members of the MTCR cannot expect India to adhere to a regime which the creators themselves violate.

However, Indian decision-makers are not oblivious to proliferation problems or the dangers inherent in the transfer of military technologies to certain states. As Dr. Arunachalam, the former Scientific Adviser to the Indian Defense Minister, says, it is understood that "vulnerability of technology in the wrong hands and the need to protect technology so that it doesn't lead to further escalation of the arms buildup are vital concerns to the US political leaders." What is also implied and must be understood is that it is of equal concern to Indians too.

India has participated actively and constructively in the negotiations putting forward a number of proposals, consistent with the mandate adopted by the Conference on Disarmament. These proposals were aimed at ensuring that the CTBT must be a truly comprehensive treaty, that is, a treaty which bans all nuclear testing without leaving any loopholes that would permit nuclear weapon states to continue refining and developing their nuclear arsenals at their test sites and in their laboratories. These proposals have underscored the importance of placing the CTBT in a disarmament frame work, as part of a step-by-step process aimed at achieving complete elimination of all nuclear weapons within a time-bound framework.

However, today, India feels that the draft of the CTBT so far remains very narrow and does not fulfill the mandated requirement of a comprehensive ban. Weak and woefully inadequate preambular references to nuclear disarmament such as those contained in Working Paper 330 cannot meet Indian concerns (Ghose, 1996). One cannot escape the conclusion that the nuclear weapon states are determined to continue to rely on nuclear weapons for their security and visualize the CTBT not as a serious disarmament measure but merely as an instrument against horizontal proliferation.

R. J. Augustus

Clandestine transfers of nuclear weapon technology, a phenomenon which has caused concern also in the subcontinent, attests to the necessity of pursuing the objective of elimination of nuclear weapons in the unambiguous format proposed by India. India has refused to sign the CTBT as it feels that the draft is shaped more by the technological preferences of the nuclear weapon states rather than the imperatives of nuclear disarmament. India remains convinced that a complete elimination of nuclear weapons will enhance global security and therefore cannot be expected to accept the treaty in its present form. India was forced by strategic considerations to reassess its past decision not to conduct any further nuclear tests after its 1974 peaceful nuclear explosion (Koch, 1996). This difference in perception with industrialized countries also has its penalties again in terms of tighter export controls, inordinate delays and restriction on technology transfer for even the simplest of technology requirements that India might have

So India's response in summary, to these regimes and the controls they imply can be evaluated through an example. In May 1997, the United States Department of Commerce put down a number or Indian organizations on its entities list . The Bharat Electronics Ltd. (BEL), Bhaba Atomic Research Center (BARC), Indira Gandhi Center for Atomic Research, Indian Rare Earths Ltd. and the Indian Institute of Science were all accused of being involved in the missile and nuclear proliferation. The appearance of these names on the entity list meant that every single item that firms from the United States wished to export to Indian agencies would need an export license whether or not the items were on control lists. The entity list also include a few names from China, Israel and Pakistan. But then considering the activities of these three countries, it appears that India does not have the same clout with the US as the others.

In any case, these new restrictions would have a wide effect on Indian research and development. There would be a limitation on the availability of dual use items and even items meant purely for commercial applications. There could be delays and cost overruns in the defense programs. India's response would have to be indigenous development and second sourcing and the penalty would be the associated increase in the R&D budget. India is determined to continue its R&D to prove that free R&D must be available globally. India would also have to export some of the indigenously developed technologies, but again as it always does, in keeping with the international norms. None of the DRDO labs or establishments have been labeled as proliferators or are they on the entity list. However a number of export license denials to US firms for technology to these labs is proof enough that the export controls are being applied. Apart from building up India's own R&D capability, the self-reliance drive is also to ensure against technology cutoff from the United States and other western countries. Although DRDO has worked towards this self-reliance, it's labs and establishments still depend on components and subsystems from abroad. While none of the current projects will be greatly affected, future large scale programs might run into rough weather. Developing countries like India should know that detailed knowledge of United States export control laws and knowing the loopholes, if any, in these laws is a good academic exercise but does not help the government or the labs they serve if western countries just decide not to export certain items. India for one will have to start a government to government dialogue and try and convince the US that the blanket application of export controls is not logical.

In Defense Of My Country

8
Indo-US High Tech Transfer

There was a mutual feeling of distrust between the US and India during the 1980's. In the United States during this time there were several in the administration, even in the Defense Department, who were of the opinion that dealing with India was risky because of the Soviet connection. Any interaction with Indians should be avoided or at best very carefully handled. It was obvious that no one really wanted to understand why India and the Soviet Union were close. In Indian society too, there were pockets that felt strongly about working with the Americans. If things went wrong for India, there were always veiled comments on 'the foreign hand', whether they meant America or not. Here too, no one wanted to reach out and make an attempt to bridge the gap. After all as we see even today, personal equations and personal opinions carry enormous sway even in a country's foreign policy.

The defense industrial collaboration between India and the former Soviet Union has always been an irritant in US-India bilateral relations. Nevertheless, it did not totally restrain the two countries from developing bilateral technology cooperation in other areas of interest. It became possible for normal government-to-government cooperation in areas of commercial technology with both governments agreeing to protect sensitive dual-use technology. The US felt India's acquiring high technology goods from the US firms centered on the danger of permitting the transfer of sophisticated technology to India when it was believed that the technology would in turn be automatically transferred to the Soviet Union. On the Indian side the concerns had to do with the thought of US controls on technology being not only offensive but also a needless requirement.

Prime Minister Indira Gandhi's 1982 meeting with President Reagan led to what was called a "blue ribbon panel" coming to India, which in turn led to the Science and Technology Initiative (Barnes, 1993). The bilateral agreements of 1984 and 1987 have demonstrated that in spite of other bilateral irritations, some amount of headway in advanced industrial collaboration could be made. Protracted but very serious negotiations eventually led to the conclusion of a 'fair and feasible' technology transfer agreement. It is worth noting here that the Indo-US understanding on technology transfer has run into some complications over the question of technologies that the US thinks may help other countries develop an intercontinental missile capability. While the end of the Soviet state in 1991 removed an important irritant in bilateral relations pertaining to technology transfer, it has not ended the obstacles to improved defense industrial collaboration. Technology transfer liberalization to India is constrained by some formal political obstacles such as India's status with respect to the two key counter-proliferation conventions and India's status as a non-signatory of a bilateral General Security of Military Information Agreement (GSOMIA) and the more recent CTBT. The coupling of these agreements and arms transfer policy has been extensively reported to the US Congress in recent years and has become institutionalized in arms transfer policy documents such as the Conventional Arms Transfer (CAT) policy document and the associated Presidential Decision Directive (PDD).

The significance of these issues is that India finds herself debarred from certain cliques of nations to whom arms may be transferred without too many encumbrances. If India is the end user, the US seems to play it strictly by the book (DODD 2040.2) and the result is a virtual case-by-case management of individual arms and technology transfer requirements. A case-by-case management of this sort with patient explanation from the Indian side and a patient listening from the American side can still resolve most outstanding technology transfer issues. Players on both sides of the court keep changing and with their changing perceptions change, playing by the book with no regard to larger ramifications becomes the theme of the day and problems arise. So, in the long-run with so large and diverse a consumer as well as an indigenous developer and producer of advanced technology as India, this is not a satisfactory state of affairs for the two bureaucracies.

It is necessary therefore that one has a general understanding of what exactly is meant by technology transfer, the US policy on technology transfer and the legislative clearances that have been passed in this regard. It is also important to understand the technology transfer process and the opportunities and spin-offs. Yet, there are issues and barriers to technology transfer and one has to analyze things on a global scale in-order to come up with a working technology transfer model. Indo-US defense technology transfer then would be another working example of this model.

R. J. Augustus

Technology Transfer

The United States feels very strongly that its national security is dependent upon its continued ability to offset, with superior technological capability, the greater numbers of systems fielded by other countries. To attain this objective, defense-related technology is being managed as a valuable and limited resource. However countries actively participating in international programs learn that the participation inevitably leads to transfer of technology from one country to another. It is for reasons of this immigration of technology that the US is becoming more and more concerned that it is losing its lead in technology. So it is no surprise that control of technology transfer in the US has been given increased emphasis over the last five years.

Technology Transfer can be defined as the set of business relationships by which technology developed in one place or for one purpose by one organization, is turned into a commercial product or process, usually by another organization. By this definition, defense conversion and dual-use are special cases of technology transfer. Of course, companies also license out technology they choose not to develop or manufacture themselves. The transfer of emerging technology may take on different connotations if it is an international border or is moving from one industry to another, or is informally passed from lab to a production unit within a company. To those working the technology transfer problem, similarities and differences are equally important. Miller states that technology and its transfer also have sociological dimensions. He says that 'the characteristics and appropriateness are a function of the political setting in which it must operate, the organizational styles that must produce and use it and the techno-economic context out of which it emerges' (Miller, 1993).

The key element in Technology Transfer is personal involvement in the interactions It is years of experience of an individual that gets translated into working solutions. Some even define technology transfer as a body contact sport in the sense that it requires a proactive approach to meet other government and industrial representatives or representatives from other countries for the purpose of promoting their use of technology developed at the laboratory. It is a combination of looking into one's own environment to get internal research personnel committed to the process, and looking outwards to encourage other companies or even foreign governments to use the technology. Figure 7 shows some of these elements that would definitely depend on person to person interaction.

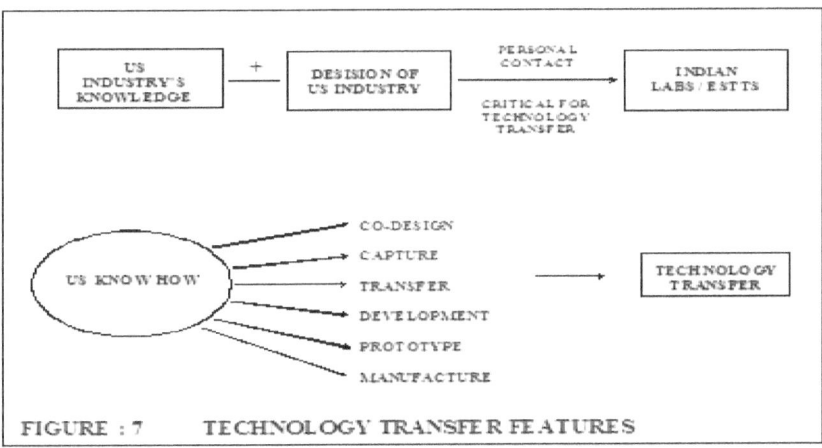

FIGURE : 7   TECHNOLOGY TRANSFER FEATURES

In the Indo-US scenario presently of interest to us, technology transfer is more the business of transferring technology originally developed at the US government laboratories or the US private industry to Indian R&D laboratories and establishments rather than private industry. The process of transfer and the results either economic or technical goes through a cycle, schematically given in Figure 8. This includes the R&D project, the invention, assessment of the invention, protection of the Intellectual property, marketing, development, licensing and revenue from the actual transfer of the technology.

The Wright Patterson Laboratory defines Technology Transfer in more stricter terms which states : "Oral or written information or data; hardware; personnel services, facilities, equipment; or other resources relating to scientific or technological

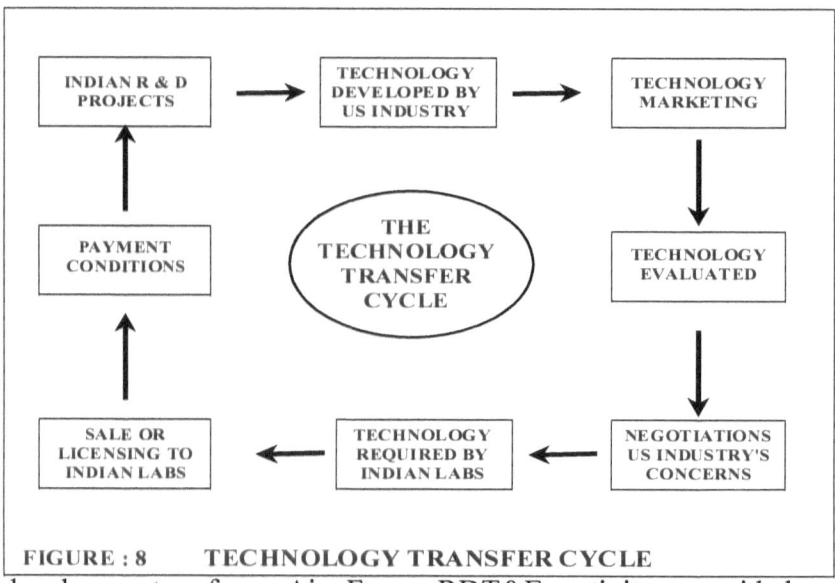

FIGURE : 8   TECHNOLOGY TRANSFER CYCLE

developments of an Air Force RDT&E activity, provided or disclosed by any means to another Federal agency; a state or local government; an industrial organization (including corporation, partnership, limited partnership, or industrial development organization; public or private foundation; nonprofit organization (including a university); or other person to enhance or promote technological or industrial innovation for a commercial or public purpose. Also includes oral or written information or data; hardware; personnel services, facilities, equipment; or other resources received by the Air Force from any of the above sources." (AFPD, 1993).

International Technology transfer has become increasingly important in recent years as a result of the perception that industrial competitiveness of several industrialized countries is eroding and developing countries need to also address this issue. International technology transfer can generally be defined as the process by which technical knowledge, information, ideas, services, inventions, and products developed in one country by any one organization, in one area, or for one purpose is applied and utilized in another country, in any organization, in another area, or for another purpose. Activities related to the transfer of technology are designed to strengthen the industrial base and hence promote the economic well-being of all nations involved. Technology transfer within the state is usually between the government and the industry, but recently, universities and other research institutions have begun to play a significant role as well.

There are certain Technology Transfer Drivers that influence the transfer of Technology as outlined in Figure 9. In western countries, most Technology Transfer efforts conducted by government agencies have been oriented toward "technology push," where a market application is sought only after development of the technology is well under way or completed.

The present trend of thinking is to improve technology transfer by encouraging more "market pull," where the actual efforts to develop technology are more focused on possible

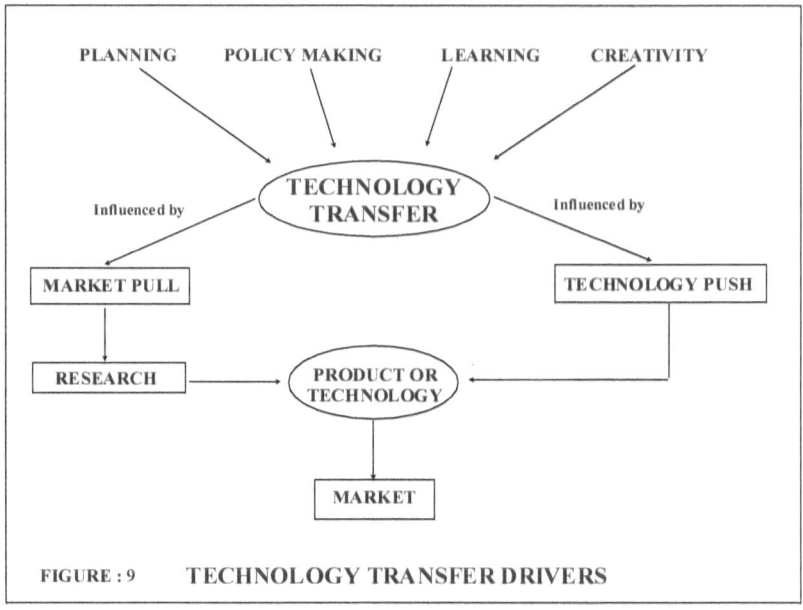

FIGURE : 9   TECHNOLOGY TRANSFER DRIVERS

technology transfer applications as well. "Market pull" and "Technology push" should be regarded as complementary components of increasing innovation. Technology Push provides the underlying base of science and technology from which innovations can flow while Market Pull provides the market need that is necessary for an invention to find application. A strong underlying base in science and technology and attention to market needs are necessary components of successful innovation.

Many governments have recognized that technology transfer should be central, but are concerned that insufficient attention has been given to a critical analysis of the underlying issues in technology transfer. The basic premise is that the decision to adopt technology is influenced by the technology itself. The implementation of technology will not be successful if the technology fails to live up to the expectations of the eventual users. Pollard suggests that some of the key points regarding technology transfer issues that deserve fuller attention are: the nature of the relationship between developing and developed countries that is being fostered by technology transfer; the full assessment of the implications for developing countries of the adoption of new technologies and a critical examination of the prevailing assumption that the industrial nations have a monopoly on environmentally sound technologies (Pollard, 1991).

US Congress Legislation On Technology Transfer

The desire for US federal laboratories to spin off technology to the private sector began with a legislation passed by US Congress in 1980. A significant portion of research and development in the United States occurs in the federal laboratories. Consequently, it was the Congress that was the motivating force that allowed commercial access to government technology. This government technology not only had the potential to contribute to the US economy, but also to provide an edge to American international competitiveness. Over the years, US Congress has continued to strengthen technology transfer legislation and this has provided the foundation for President Clinton's initiatives to stimulate the US economy and improve international competitiveness through technology transfer. Looking at it from purely the defense angle, one must bear in mind generally the US Department of Defense will not transgress its technology transfer policy.

Transfer of technology does not happen overnight as no one is willing to part with the 'crown jewels' – that which they have struggled to achieve and learn unless they have a very good motivation to do so. Within the US, several acts of congress had to be brought into force before there were working equations for the transfer of technology to occur. The Stevenson-Wydler Technology Innovation Act and the Federal Technology Transfer Act of 1986 are the two major events. The Stevenson-Wydler Technology Innovation Act (P.L. 96-480) made technology transfer part of the mission of every federal laboratory and established within each laboratory an Office of Research & Technology Applications. The Federal Technology Transfer Act of 1986 (P.L. 99-502), which amended Stevenson-Wydler, provided significant new authorities for Air Force laboratories to enter into Cooperative Research and Development Agreements (CRDAs) with private companies as well as public and nonprofit organizations and to negotiate license agreements of intellectual property on behalf of the government. One can list some of the technology transfer legislation milestones as follows:

1. Stevenson-Wydler Technology Innovation Act of 1980 (P.L. 96-480)
2. Bayh-Dole Act of 1980 (P.L. 96-517)
3. Federal Technology Transfer Act of 1986 (P.L. 99- 502)
4. Executive Orders 12591 AND 12618 (1987)
5. National Competitiveness Technology Transfer Act of 1989 (P.L. 101-189)
6. American Technology Preeminence Act of 1991 (P.L. 102-245)

If such was the US domestic case for moving technology from the government labs to the commercial sector, it is not surprising that third world countries have such an arduous task in prying technology for themselves from the US. On the international platform, those who have work the technology transfer game, sometimes feel that it takes on the appearance of a myth or even an illusion and there is a feeling that the industrial nations have no intention of making available high technologies to the developing world due to military considerations. "The South has no choice but to develop these technologies on its own through greater reliance on South-South cooperation" (Elmandjra, 1991) and even when a transfer occurs, quite often the question raised is whether what is being transferred is actually technology or just techniques.

DOD Policy On Technology Transfer

The primary policy governing the process of technology transfer in the US is contained in the Department of Defense Directive (DODD) 2040.2, "International Transfer of Technology, Goods, Services and Munitions" issued on 17 January 1984. This directive for the first time institutionalized responsibilities for the security of technology within US department of defense and established a working relationship among the offices of the Under Secretary for Policy, the Under Secretary for Research and Engineering, the Organization of the Joint Chiefs of Staff, the Services, and the Defense agencies. The directive DODD 5105.51 further tuned the process by establishing the Defense Technology Security Administration (DTSA). The effect of this was to enable the DOD to formulate for themselves a more efficient, predictable, and transparent procedures for reviewing export licenses even though to an outsider the entire export license process is still a labyrinth.

R. J. Augustus

The technology transfer policy is set forth in DODD 2040.2 in very clear terms. The basic references for this has been (a) Public Law 96-72, "The Export Administration Act of 1979," as amended, (b) Public Law 94-329, "The Arms Export Control Act," as amended and (c) National Security Decision Directive Number 5, "Conventional Arms Transfer Policy," July 8, 1981. The policy declares that it shall "treat defense-related technology as a valuable, limited national security resource, to be husbanded and invested in pursuit of national security objectives." The DOD also says that since the US values international commerce, they would apply export controls in a way that 'minimally interferes with the conduct of legitimate trade and scientific endeavor'. With this in mind there are seven guide lines laid down . They are:

1. The transfer of commodities like technology, goods, services, and munitions must be consistent with US foreign policy and national security objectives.
2. Control the export of commodities that can contribute to a country's military potential or prove detrimental to US security interests.
3. limit transfer of advanced design and manufacturing know-how
4. Share military technology only with allies and other nations that cooperate effectively in safeguarding these commodities from transfer to nations not friendly to the US.
5. Make sure that rapidly emerging and changing military technologies do not reach
   potential adversaries before adequate safeguards can be implemented.
6. Strengthen foreign procedures to protect these commodities by cooperation and

7. Work towards reciprocity of technology transfer.

The country desirous of receiving such defense technology has to qualify to certain US requirements. This country would have to maintain control and restrict the transfer or export of US defense commodities to other nations who may misuse it. Written permission from the US is required before re-export and the US must be promptly informed of any known or suspected US unapproved transfers. These countries should also ensure that they do not trade in non-commodities that may threaten US security.

Looked at from the strict policy angle, there should be no room for countries to receive technology that the US does not want to export, provided the policy is followed to the letter. However in today's scenario, there are concessions made and compromises are reached in the rough and tumble of trade, politics and defense relations.

The Technology Transfer Process

Technology Transfer appears to be a simple process of communicating one's ideas and experiences whether in the national or international context. However, detailed study and analysis reveal that a predictable learning pattern has to be gone through. In this learning curve, one gets to first understand the technology which in many cases may differ from the theory, the analysis of how this technology may be used to solve a problem and finally, the actual application of the technology to solve the problem. Scientists and engineers can influence this learning pattern, once they obtain a basic understanding of the technology transfer process.

R. J. Augustus

The actual technology transfer process from the Indian perspective, forms one major cog in a much larger acquisition chain of activities with certain technology drivers acting as catalysts between technology requirements and the technology transfer. Figure 10 starts with the project requirement from either the defense services or as a pure R&D project. Once the project proposal is made, sanctioned and technology specifications and requirements are known, one has then to identify the technology components that have to be acquired. The required technology could be either new technology, derivable technology or resident technology. Based on this, the availability within the laboratory, within or out of the country has to be ascertained. Once the sources of technology have been identified and contacted, one goes through the rituals of requesting for a technical proposal and in some cases this would be accompanied by a request for the

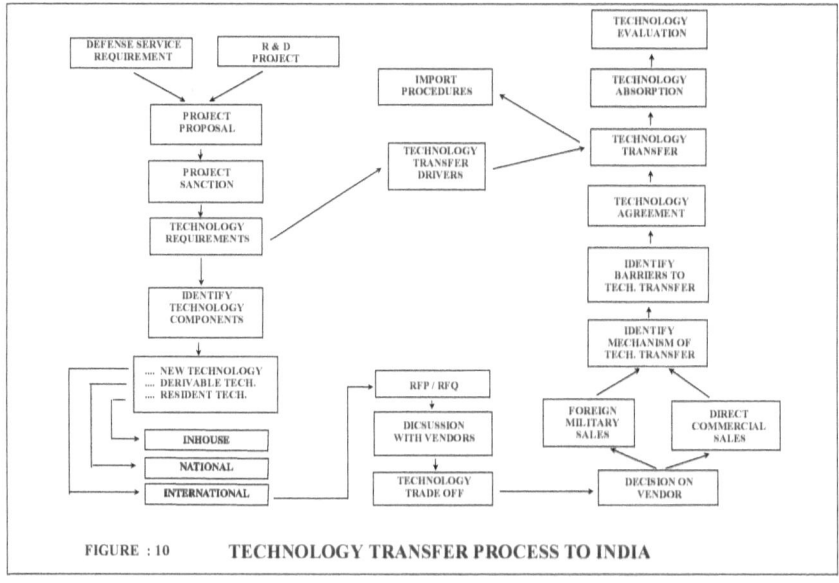

FIGURE : 10   TECHNOLOGY TRANSFER PROCESS TO INDIA

quotation. The receipt of the technical proposals from interested parties would then trigger a technology-cost trade-off study in order to short list the suppliers. Unfortunately what happens in most government trade-off studies is that there is a tendency for going in for the lowest quotation or in some high handed cases decide on a vendor with total disregard for the study. In any case what happens next, at least in US circles, is to decide whether one goes in for a Foreign Military Sales (FMS) or a Direct Commercial Sales (DCS). What follows then is a process that includes identification of the transfer mechanism, identifying barriers to the transfer, tuning the technology agreement, the actual transfer to the recipient country, technology absorption by the recipient and finally the technology evaluation by the testing agency, the user and the government.

It is obvious that for the technology transfer process to be a success, a congenial and an apt environment must be provided. Figure 11 represents a schematic view of the environment and the participating groups. The equipment, materials, methodology and above all the people participating are part of the recipe for success. As mentioned earlier, it is the working group on the supplier's side the recipient's side who are the fundamental cogs in the transfer process for setting the pace and quality of the transfer. Some of the key participants in the process are schematically represented in Figure 12. Starting with the researchers and engineers right through to the Government policy makers in the US and India, every single individual has a major role to play in the success of the transfer.

There are certain indicators in the technology transfer process that represent signposts to those involved in the transfer. Figure 13 represents the technology transfer indicators and shows what occurs in each phase. These six stages described by Risdon

R. J. Augustus

(Risdon,1992) may be considered essential steps in the context of International Technology Transfer and perhaps specifically in the Indo-US context, even though there is a basic difference in the way India and the US generally go through the different phases.

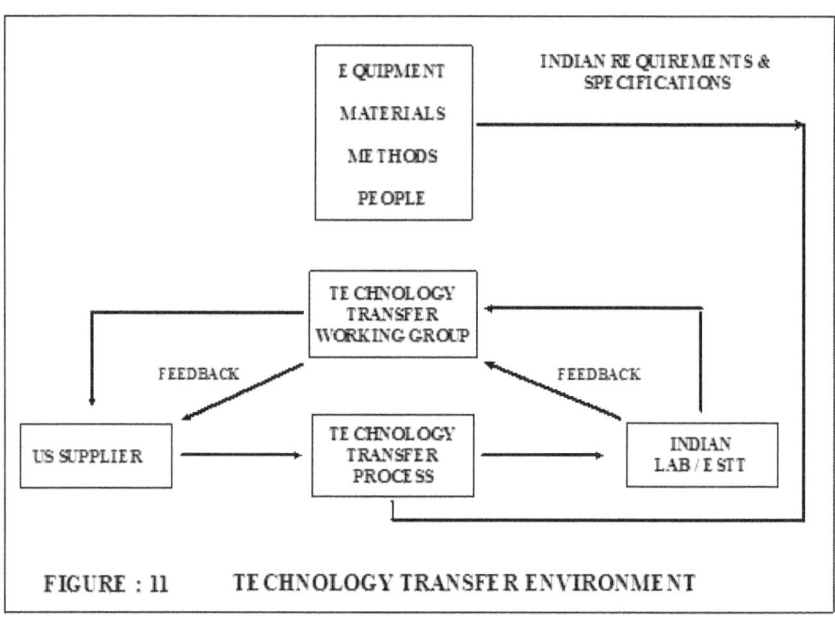

FIGURE : 11   TECHNOLOGY TRANSFER ENVIRONMENT

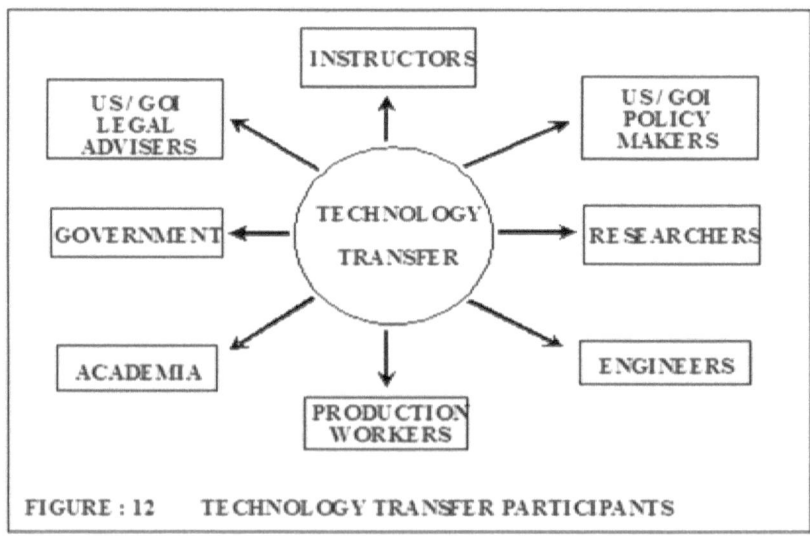

FIGURE : 12   TECHNOLOGY TRANSFER PARTICIPANTS

The six phases comprise of the conceptual phase, the feasibility phase, the recipient phase, the commercial phase, the application phase and the evaluation phase. It is not essential that there is a clear demarcation between each phase for it depends on the type of transfer one has in mind and also the scope of work for each interested party. A clarification of the elements involved is essential to understanding the contribution of these stages to the transfer process.

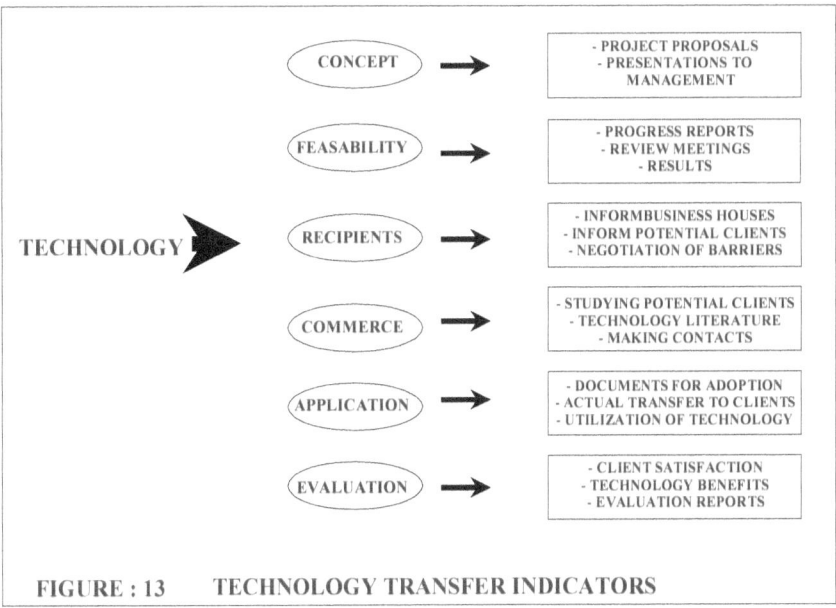

FIGURE : 13   TECHNOLOGY TRANSFER INDICATORS

Conceptual Phase : The technology conceptual phase starts with the scientist or engineer who has a problem to solve or a product to design based on customer specifications. This leads on to the exchange of information with similar persons in other labs or countries, ideas of how the technology can be used to solve a particular problem or improve a situation in a research priority area. Some are of the opinion that the technology transfer starts right here, for this phase represents the exchange of information which takes place between scientists, engineers, colleagues and administrators to advance ideas on the application of science and which is beneficial to others. So logically, any assistance which is given to others, thereby facilitating their work has commenced here and so it is the start of the technology transfer process.

Feasibility Phase : Once the technology has been conceptualized, the feasibility phase is represented by the scientist or engineer by first conducting research which generates data in support of the underlying theory. These results are then communicated to colleagues, peers and administrators quite often in the form of reports which act as indicators to signal that the research efforts have been a success and therefore the pursuit of this technology is feasible.

Recipient Phase : During the third phase of the process, decisions need to be made concerning who needs the technology. The people involved in this phase would be scientists, engineers and marketing personnel. These specialists need to be aware of factors such as cost, convenience, environment for the technology, strategy of the recipient, present and future requirements of the recipient, political climate in the recipient country and so on, which influence recipient's acceptance or rejection of new technology. In addition, there is also the crucial question of the recipients qualifications to receive the technology according to national and international export laws. Software and programs for checking this online are available and is encouraged by the US administration.

Commercial Phase : The commercial phase of the process is concerned with disseminating the mature technology to the outside world. In the case of international clientele, approaches are made directly or through representatives like the service attachés or technology advisers in the country where sales are sought. Key actions for technical liaison involve the talents of scientists, business leaders and marketing specialists to educate recipients to the various benefits of the new technology. It is in this phase that research, engineering and sales personnel develop a knowledge of future customers and start to target them. Knowing where and how the potential clients look for available technology will help the vendor team in covering those areas. In this phase the vendor also needs to gain approval of his government for exporting the technology.

Application Phase : The application phase concerns the understanding of the actual integration of the technology in the recipients environment. This involves the installation, educating the recipient about the utilization of the technology and monitoring his application of the technology to his needs. This is a very useful phase for the vendor as he gains insight into customer behavior and can utilize it again in improving his concepts in the first phase. If it is a joint development and integration of technology, care must be taken to keep channels of communication open with teams resident in each other's environment. Individual teams trying to develop and integrate technology while staying miles apart may be a disaster in incubation. Indian R&D has gained a lot of experience on this score – sometimes to its grief.

Evaluation Phase : This phase of the technology transfer process documents the success or lack of success of the adopted technology . This applies to both the vendor and the recipient. Key actions for the evaluation phase are to establish assessment criteria for identifying and evaluating socio-economic and environmental benefits or harm. Assessing technology transfer effectiveness generally requires an in-depth study of the quantitative and qualitative benefits for both parties.

Technology Transfer Mechanisms

In the US, mechanisms for technology transfer have been provided by Public Law 99-502 and the Technology Transfer Act of 1986 which requires federal laboratories to actively seek opportunities to transfer technology to industry, universities, and state and local governments. The transfer of technology generally falls into six categories such as 1. Technology assistance, 2. Patent licensing, 3. Cooperative research and development agreements, 4. Education Partnerships, 5. Cooperative Agreements and 6. Grants.

Under the technology assistance method, technical assistance is requested to help in solving a technical problem is usually made to the Technology Transfer Office by anyone interested in the technology. The technology transfer office will then introduce the two parties. Under patent licensing, those interested may seek permission to utilize the technology for development or licensed production. The goal of the Cooperative Research and Development Agreements (CRDAs) method is to conduct joint research in a technical area of mutual interest while ensuring that both parties have equal access to the results. While the government laboratory may contribute people, equipment and facilities to the collaborative effort under CRDA, the interested party may contribute the same as well as money if required. Under the nonprofit Education Partnership method, a government laboratory may loan or transfer equipment and laboratory personnel to assist a university or nonprofit organization in the development of programs or courses. The Cooperative Agreements (CAs) method is usually competitive and may require cost matching of resources. It is used where there will be significant involvement between the Government and the recipient during performance. The sanctioning of grants is the sixth method which a government lab might share with educational, nonprofit, and state and local governments only.

US agencies can transfer technology to Indian DRDO labs and establishments in several ways. Some types of defense technology transfer are more applicable in direct lab to lab cooperation or industry to lab or to university while others have to be worked out between the two governments. The more common types are: 1. information exchange; 2. consulting services; 3. use of US government facilities for research, manufacturing, repair, or testing ; 4. exchange programs for transfer of people to or from an establishment ; 5. cost-shared contracts where the defense services, private industry, academia, or other government agencies share the cost of a joint effort; 6. License agreements where the defense services transfer or private sector are given permission to use the technology; 7. Participation by defense employees in professional organizations and societies; 8. participation in consortia where scientists and engineers from defense establishments, private industry, and academia work together on R&D programs; 9. participation and funding by businesses in Indian research programs, academia or private industry and 11. technical assistance provided to Indian universities, colleges, and high schools to avail of manpower and research capabilities.

In the numerous interactions that the Indian DRDO has had with US industries, government and academia, the transfer of technology has broadly followed the same methodology even if the content has varied on a case by case basis. Figure 14 represents some of these methods used in Indo-US interactions. The US is one country where copious amount of information is available from a plethora of sources. However in all the interactions, the US government, US industry and any other agency including universities, extreme caution is taken in protecting themselves and not disclose information prematurely if the research may result in a patent application or if other proprietary information is involved. Conferences and publications at universities including sabbaticals and presentations at professional and technical conferences on R&D results are true and tried mechanisms of technology transfer. Conference presentations published for distribution to conference attendees often reach individuals who may not have attended the conference. Technical consulting is also useful though it must be carefully spelled out with regard to scope of work and time schedules, to ensure there are neither conflicts of interest, nor potential intellectual property concerns. Exchange programs to exchange expertise and information, provide for a transfer of personnel either from lab to lab or lab to industry domestically or internationally. In any case, contracts and patent rights usually go hand in hand and the preparation and execution of these agreements by both parties must be done as per law and with a code of ethics.

Indian interactions with US defense industry has driven home several hard truths. Because of US export controls, Indian labs have learnt that consultants need not give you solutions. At most they may just point the recipient in the right direction. If they cross the drawn line they are bound to be hauled up as in the case of Loral for allegedly offering solutions to the Chinese on their

In Defense Of My Country

launch failure. Shared research will be of benefit to both parties, but one should not be surprised to find the US industry or the US government holding on to 'proprietary' information for a certain period of time. It is difficult for recipient countries to accept these peculiarities in technology transfer.

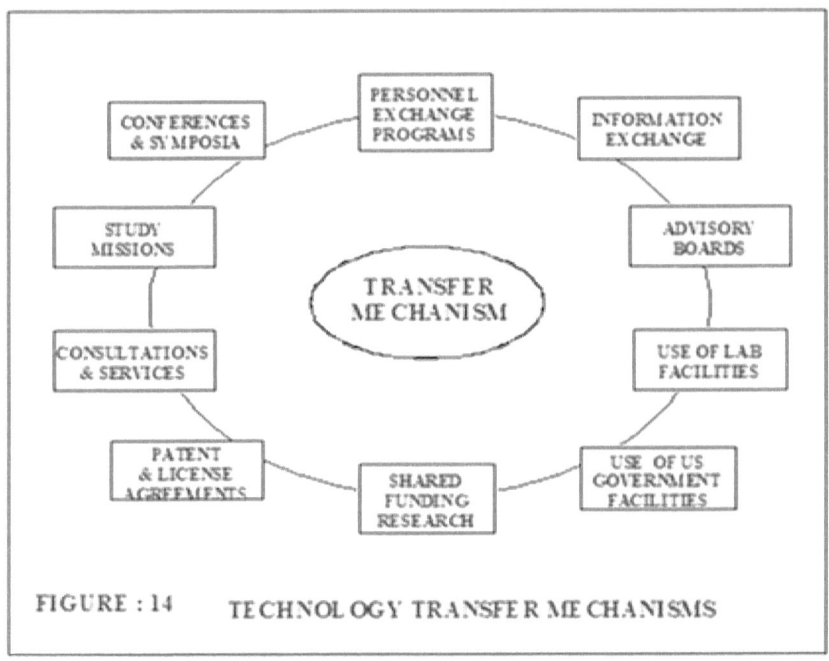

FIGURE : 14   TECHNOLOGY TRANSFER MECHANISMS

Today from the Indian R&D point of view, there is a need for the Program Manager to (1) understand the process by which technology can flow from US firms to India; and (2) recognize the need for adequate assessment of all technology required for program implementation in-order to permit decision makers to assess the DRDO investment against the desired technology and program objectives. Before any agreement is made or negotiation begins to transfer technology, the US must, based on its DODD policy, decide (1) whether or not the technological information should be shared, and if so with whom; and (2) how the information is to be transferred and how collaboratively developed technology should be protected.

Based on experience that Indian DRDO laboratories have had in technology transfer outside the country, within the country and the acquisition and absorption of this technology, an attempt has been made to put down these interactions in the form of flow diagrams. Though these figures lean heavily toward the specific interactions of the Indian Defence Research and Development Organization, they can also be adapted quite easily to the needs of any other developing country.

The Trade-Off

When one first starts to logically think about technology transfer, the first thing that comes to mind is to identify what technology is required to be transferred. Whether the desired technology is critical technology or conventional technology, the performance of the technology has to be specified and the cost involved has to be studied. The performance would depend on the hardware, the software and any training that might be given by the supplier. The cost involved can be broken down into R&D costs, capital costs and administration costs. Once these crucial areas have been studied, the laboratory or establishment can then start shopping around for the technology utilizing a simple trade off practice. The suppliers could be foreign countries or even domestic suppliers. Figure 15 represents the tradeoff studies carried out by a DRDO lab when dealing with a foreign supplier.

The supplier is chosen based on a tradeoff between technology and cost, taking into consideration the recipient's priorities. At this stage, the major consideration is whether the firm does have the license to export or is confident of receiving the license. Quite often due to the desire to 'clinch the deal', the supplier assures the recipient that they would receive the export license. However after months of work, opening of letters of credit and other preparations, the recipient is informed of the US government's intent to deny the export license or of its denial. So the recipient would now have to convince the US administration to issue the export license or look for an alternate source. Hence the decision to discuss this issue up front must be appreciated. Recipient education or training, the technology transfer plan and firm transfer milestones are important considerations in the trade off which leads to the final choice of a vendor. With the choice of vendor confirmed, both parties can then enter into formal negotiations on the technology transfer articles of agreement.

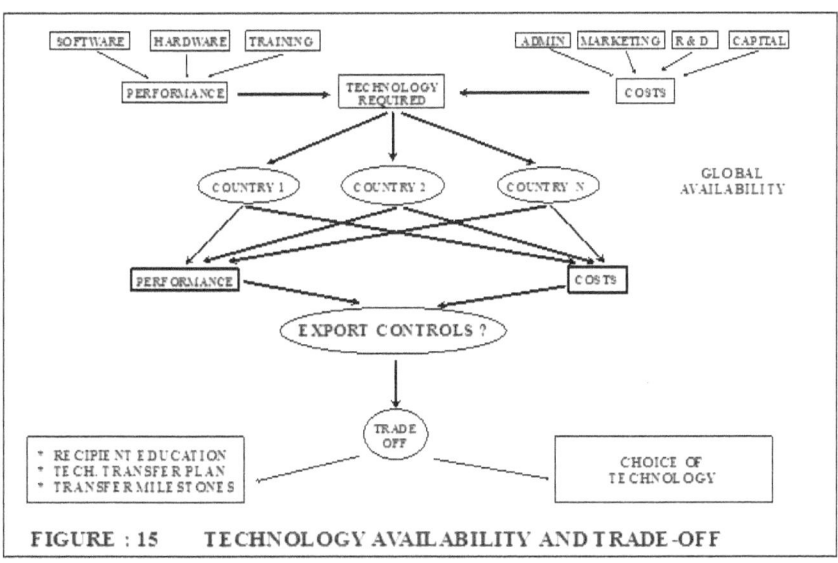

FIGURE : 15   TECHNOLOGY AVAILABILITY AND TRADE-OFF

R. J. Augustus

Articles of Agreement

A harmonious uncontroversial transfer of technology depends on the supplier and the recipient conforming to certain rules and regulations that must be put down as articles of agreement before the transfer. Based on the definition of the technology involved, both teams must draw up a scope of work with time frames and milestones specified. This would include the effective date, effective period and termination. The agreement on price should include the payment schedule, escalation, terms of payment that includes taxes, duties and levies. The delivery title, risk of loss, inspections and acceptance procedures, proprietary information, intellectual property rights, patents and warranties must be addressed and agreed upon. In case of delays or non-conformance of the articles of agreement, method and place of arbitration are to be fixed taking into account notices and limitations of liability. Finally, there should be an agreement on international environment and nuclear indemnification laws. Figure 16 displays some of these articles of agreement.

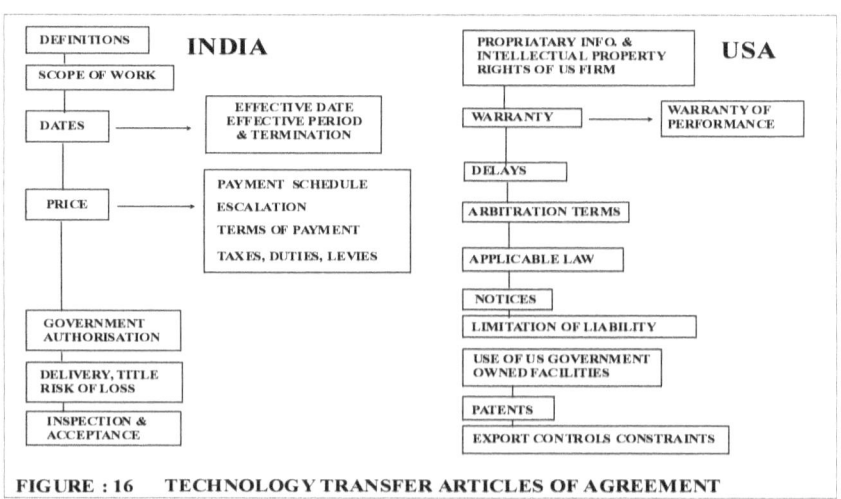

FIGURE : 16   TECHNOLOGY TRANSFER ARTICLES OF AGREEMENT

One must also understand that at times due to the anxiety of getting the project going, the recipient and the supplier do not pay much attention to the fine print. During the project these issues could always surface and create controversies that result in excess payments and project delays. There are two important points that need to be considered by foreign vendors. Because of their vast experience in working large projects and technology transfer they should not imagine that the labs and establishments in Third World countries are also completely familiar with issues involved. Secondly they should not take advantage of the naiveté of these establishments. After all, there should be a code of conduct for the US agency would not deal with the US Department of Defense in like manner.

Technology Transfer Evaluation

The evaluation of international technology transfer is no mean task since it covers a wide range of activities. As such it seems quite difficult to gauge the success of the transfer. But since governments who have invested in the transfer would like to see the results, there has to be some pre-established criteria by which some measurement can be made. As Carr (Carr,1995) says, knowledge is the coin of technology transfer and it is spread over and embedded in every aspect of technology transfer. While to some extent one can arrive at the qualitative content of technology transfer, it is extremely difficult to quantify the experience gained doing such an interaction. There is a similar problem when one wishes to gauge the economic impact of technology transfer. To measure the economic value of technology is to try and measure the economic value of knowledge. So it is unlikely that one can come up with a significant way to link inputs and mathematically compute economic benefits given the complexity of the game players and the events. The success of a project would be representative of the quantitative value of the transfer while the qualitative assessment will comprise of several factors including the comments and experiences of the individuals concerned. However, over a period of time, success stories like technology partnerships and spin-offs will yield enough information to be analyzed and put together as an indicator that can be measured and provided to policy makers. The best indicator for the success of the technology transfer would be to evaluate whether the labs to which the technology transfer was made, is self-reliant or not after a few years.

Technology Partnerships

Fred Adler (Adler, 1996) describes international cooperative research and development as a technology transfer mechanism that can reduce the development costs of weapon systems and combat support equipment and enhance interoperability among friendly countries. With most countries, India included, trying to restrain if not shrink their defense budget, it is important that every opportunity for a win-win partnership is examined. Today one sees this trend in most major aerospace design and development projects. Partnerships between governments can mutually utilize each other's resources to meet common or compatible objectives. By accessing capabilities and talent found in other countries, the recipient country can gain valuable benefits, such as reduction of their technical risk, and access to unique facilities and people with skills outside the scope of their own work-force and try to achieve vital leading edge technological capabilities. The utilization of Indian software expertise by multinationals drives home this fact.

The Indo-US interaction on the on-going Indian Light Combat Aircraft Program indicates that a successful partnership may benefit far more than the mere sum of the parts. In a successful partnership, both parties save money by doing the research together rather than separately. More importantly, both gain from access to each other's brain-power, technology, know-how, facilities, and culture. This is possible only when some basic principles are adhered to when entering into partnerships. Partnerships must directly support the country's missions and must have the potential to provide economic and other benefits to the nation. They must be formed in a fair and open manner and should use best business practices. However, efforts must be made to ensure that the results are measurable in some form.

For a developing country in particular, results must be visible and the success of every technology-related action in any field or area depends on transferring the technological results from the foreign lab or establishment into useful equipment or programs required by the recipient countries. The knowledge gained from this technical interaction must be evaluated, analyzed, and shared so that practical applications for the new knowledge can be found. Instead of resting there, the technology must be further developed and demonstrated to cater to further requirements of the user. Whether in the form of patents, information, standards, processes, or physical products, the technology from a transfer must, in the long run, be adopted by industry, consumers, and other end-users if it is enhance the recipients industrial capabilities. In addition, the increasing competition among nations in the international marketplace and the increasing requirements of one's own defense forces demands a faster technology transfer.

Spin-offs

Spin-off technology transfer can occur when technology developed by one organization, in one technical area, and usually for one purpose, is applied and used for a different purpose in a different technical area or market application than those foreseen at the time the research and development was originated. Although spin-off transfer is usually associated with the fundamental and basic science research programs within a government agency or organization, there could also be opportunities for spin-off transfer in each of the applied research and development programs. India's 'Missiles to Medicine' attempts are a good example of this.

The transfer and dissemination of useful information and knowledge is implicit in the transfer of technology. For example, for a third world country, once the initial idea of a certain requirement is formulated within an organization, the path to successful absorption and management of the technology is a long and tedious series of drawing up specifications, market surveys, negotiations, interactions in design, development, testing, and production phases and actual transfer of technology apart from the technics. This is also true of a pure research and development laboratory attempting to manufacture and market products.

Indian Acquisition Cycle

Since the focus of this technology transfer study is basically one of transfer to India, the analysis would not be complete without a look at the present acquisition and absorption apparatus in India. This is the apparatus that will be used to facilitate the transfer of technology and two flow charts (Figure 17 a; 17 b; 17c and Figure 18) have been developed to indicate the functioning of the systems. The acquisition cycle starts with requirements projected by the defense services with which the technology to be acquired can be identified. The organizations policy and technology forecast department would then hold a review into which inputs from the production department, quality assurance department, the actual user, the DRDO and the Indian representatives abroad will be funneled and used. The preliminary feasibility study by DRDO that follows decides whether the technology should be bought or developed domestically. The make or buy decision being taken, all concerned departments including the department of standardization and electronics review the defense staff requirements and the DRDO is then charged with making a detailed feasibility study using inputs from all nominated agencies.

Two separate paths are followed for the make and buy options as indicated in Figure 17 b . In the make option, model 'A' is first developed by the MOD and DRDO using inputs from the user and the testing agency. The functioning of this model dictates whether one should proceed with the 'make option' or go out and buy. Model 'B' is developed based on a decision to make. This model is then subject to technical and user trials. The decision to develop the technology in-house involves the DRDO and the production agency. If the decision has been to buy, then what follows is basically a process explained in the Trade-off earlier. In this case the initial model supplied by the vendor is used for technical trials and user trials on a no-cost basis.

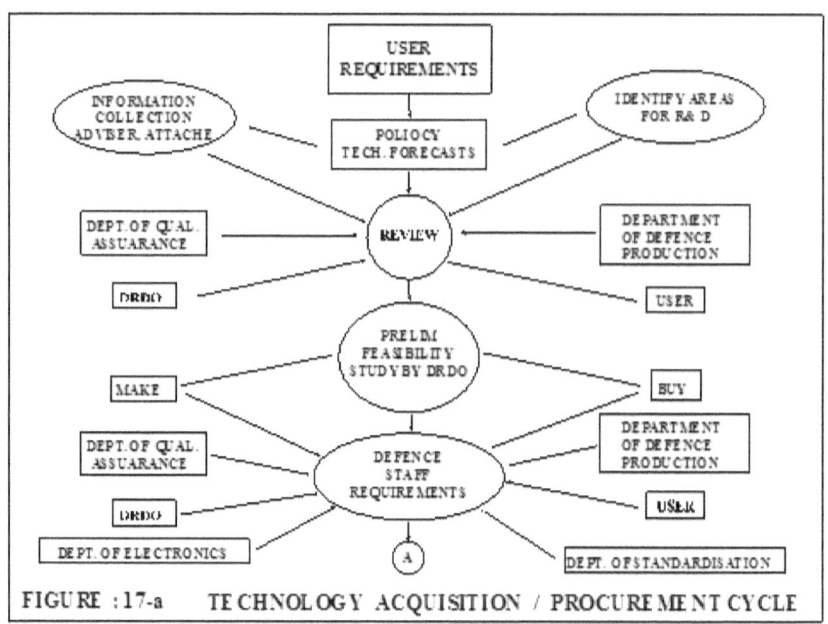

FIGURE : 17-a   TECHNOLOGY ACQUISITION / PROCUREMENT CYCLE

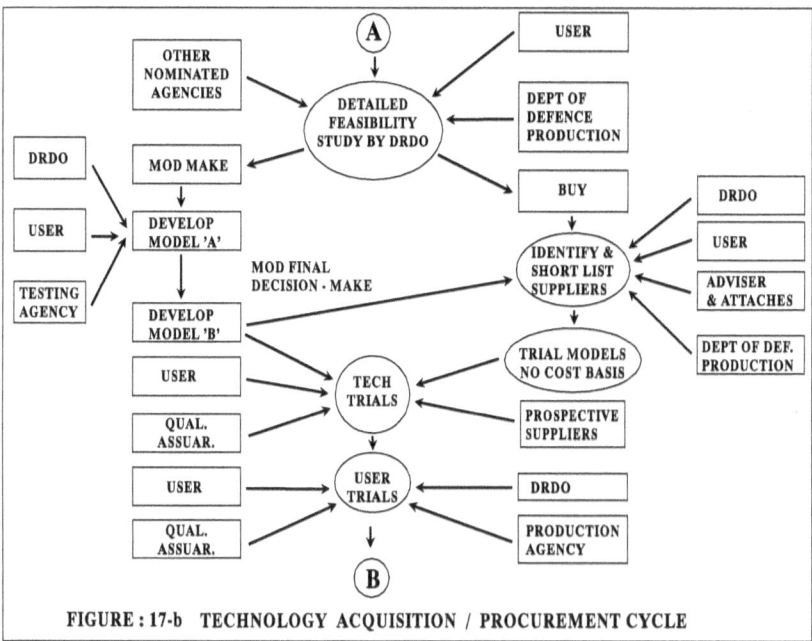

FIGURE : 17-b   TECHNOLOGY ACQUISITION / PROCUREMENT CYCLE

In Defense Of My Country

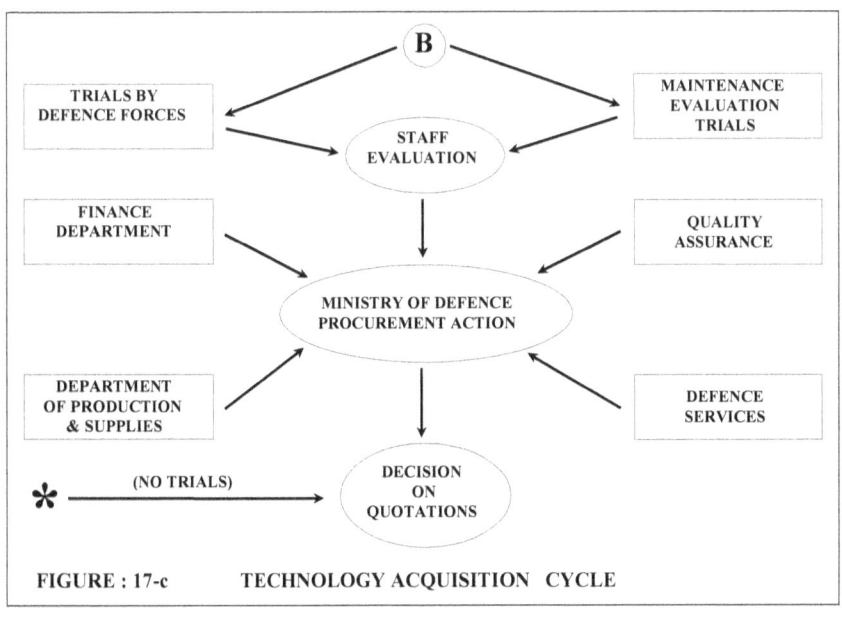

FIGURE : 17-c   TECHNOLOGY ACQUISITION CYCLE

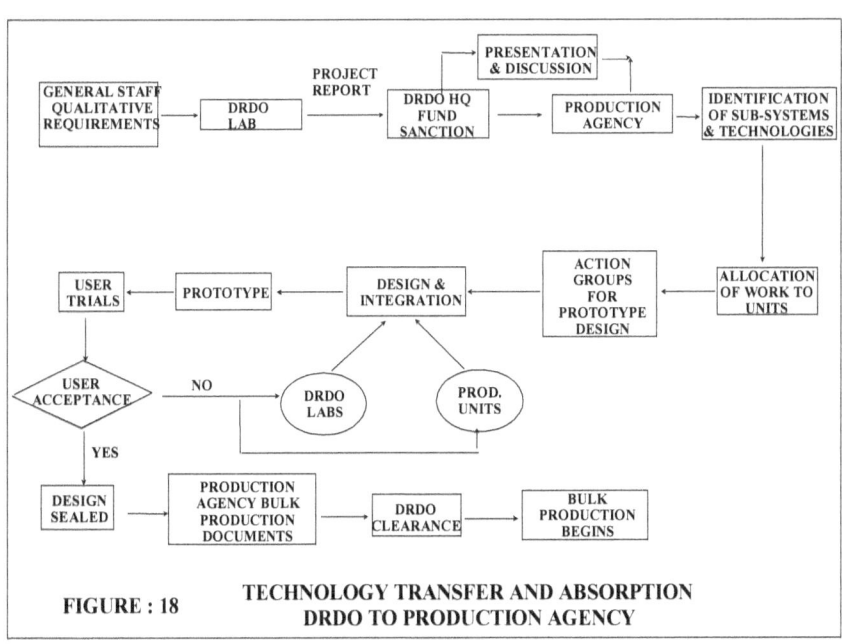

FIGURE : 18   TECHNOLOGY TRANSFER AND ABSORPTION
DRDO TO PRODUCTION AGENCY

The actual field and maintenance trials of the domestically built or bought system are then carried out for a complete staff evaluation. The department of finance, department of production and supplies, the quality assurance group and the concerned defense service decides on the manufacture or procurement action which then leads to the decision on the negotiations with the supplier. This process is represented in Figure 17c.

To complete the picture, a detailed view of the in-house interaction between the DRDO and the production agency must be studied. In Figure 18 one sees the chain of action that is responsible for the start of the project to the production. From the staff requirements, the DRDO is responsible to bring out a project proposal which will be the basis for the sanction of funds. Detailed discussions with the production agency lead to the identification of sub systems and the technologies needed. Once this is identified then the individual components of the work involved can be allocated to the different R&D and prototype units. Their interaction brings out the prototype design. With the design and integration stage complete, the prototype is then subjected to user trials. If the user is not satisfied then no acceptance is given and the prototype goes back for redesign and modification. It is here that the DRDO has quite often felt the pinch because the user comes up with specifications and a wish list that was not clearly specified during the initial staff requirements. One cannot really blame the user because during this time technology has improved and the user who is now aware of new developments would like the DRDO to integrate the new technology into the design. No foreign vendor would agree to such requests, but the DRDO strives to accommodate the user requirements thereby leading to design problems and delays. The design and development of the Indian training simulators of the 1980s is a classic example of this issue.

However, after final designs are complete, tests are carried out and the acceptance is made. The design is sealed and work starts on the detailed documentation for bulk production. After DRDO's final clearance the production agency is free to start bulk production.

Barriers To Technology Transfer

Like any other interaction, international technology transfer too can run into some sticky situations. Close attention must be paid to the barriers that can stop or delay the transfer of technology. Most developing countries appear eager to receive new technology and technology transfer is being portrayed as one of the "gifts" offered by the industrial countries in exchange for certain considerations. But some analysts feel that it is an eagerly awaited opportunity to open new markets, and to develop new licensing agreements whose profits over a long term period will outweigh the cost of any initial investment. Developing nations may become caught up in the process of negotiating the scope and generosity of agreements for the transfer of technology from the West without serious study and analysis.

Since there is so much movement towards globalization of industry after the change in the political economy of today's world, one needs to address questions that will most certainly arise regarding the policies and practices that must be followed by industrial houses and governments engaging in technology transfer. Firstly, there will be a new set of dependency equations that will come into play between the supplier and the recipient based on the transfer. Secondly, developing countries must take care to see that the introduction of new technologies does not undermine their self-reliance programs. This will put them in a spot if it results in dependence on the supplier who might not be able to live up to expectations due to the uncertainties of international politics. Thirdly, one must consider the impact of new technologies in the lab and the organization. Very often technology transfer is considered in isolation from the context or environment into which the technology is being introduced. The defense sector may not have as many problems as the commercial sector in ensuring that technology transfer does not disrupt local economy, labor or the environment, but a strictly financial analysis of technology transfer may not come up with any barriers to transfer. Despite the cautions that have been expressed concerning the prevailing patterns of technology transfer, Pollard says that it is clear that there will be significant transfer from developed to developing countries, and that there is need for close attention to the terms of the transfer (Pollard, 1991). In the vein of this argument, there is a need for developing countries to identify the conditions where a clear interest exists for technology transfer and provide suitable and acceptable justifications for adoption of the specific technology.

Guenther Cyranek, Head of the Technology and Society Unit, Gottlieb Duttweiler Institute, Rueschlikon states that as research has demonstrated, the shortage of technological, management capacities and human resources, has been a major cause for the failure of many previous efforts to nurture independence from developed nations. He says that "in the reorientation and revision of the development policies which have been followed in the 'Third World' until now, education, training and the development of human resources to integrate technology into the development process, must play a central role" (Cyranek,1991). Indian technically skilled workforce does provide a strong basis for self-reliance.

Anything that impedes communication within the lab or organization will jeopardize the successful implementation of the technology within the organization. Those using new technology must change their behavior patterns. So it cannot be expected that the recipients will respond to new technology quickly. In addition, it is human nature to resist ideas, especially those originating from outside of the organization, and this can lead to a very limited and narrow vision. A clear implication is that technology transfer requires time, patience and opportunities to experiment or become familiar with a new technology.

The adoption of new technology is also heavily influenced by the climate in the organization. If there are clear advantages and the technology does not drastically interfere with existing practices and is easier to understand, then it is accepted. An added incentive would be to have introduce technology that can be easily measured and evaluated. Management is always looking for the visible benefits of technology transfer. So this brings us to the actual barriers that can stand in the way of technology transfer.

Figure 19 represents some of the critical barriers to technology transfer. These can be separated into the technical and non-technical barriers which concern the defense industry, academia, and government. The Technical Barriers arise due to the lack of certain technical capabilities in the recipients laboratory or organization. One may wish to include the lack of standardized testing methods, standard design methodologies, reliable analytical and predictive tools, comprehensive material database, modern and cost effective manufacturing experience and facilities and so on, as factors that can impede the transfer and absorption of technology. When the non-technical barriers to technology transfer are considered, one comes up against government trade regulations, intellectual property rights, export controls, international relations and skill base of the country's workforce to name a few.

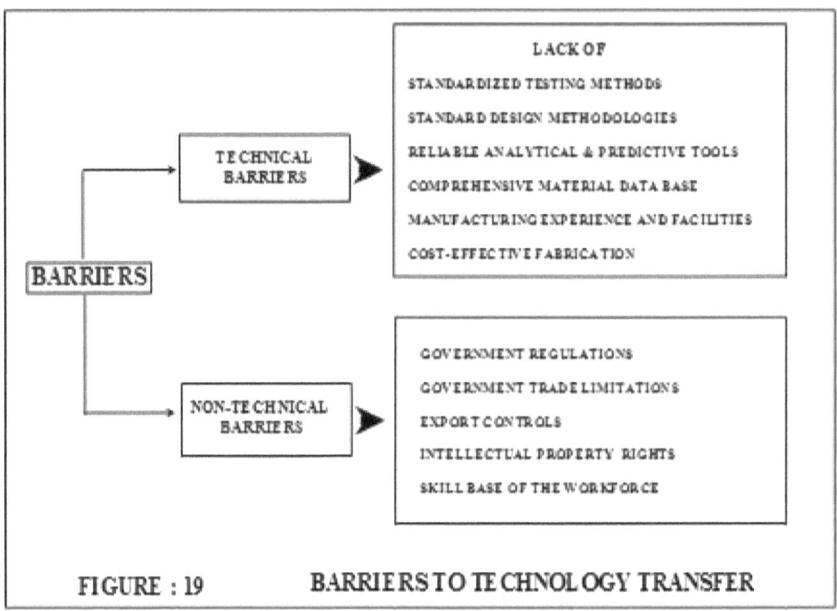

FIGURE : 19   BARRIERS TO TECHNOLOGY TRANSFER

Though in most cases all technical barriers for technology transfer can be satisfied, one is left with the more difficult non-technical barriers which are often subject to the political equations between the countries involved. It has been identified that problems arise due to poor communication of information from the technology recipient to the supplier and to the suppliers government. This is especially true while dealing with information vital to fulfill government regulations and export control requirements. So recipients are recommended to take extra care in giving all information relevant to facilitating the transfer. Sketchy or ambiguous information could for example lead to a denial of an export license or at best introduce a prolonged delay.

The Export License

When an Indian lab places an order with a US defense firm, the firm has to apply for an export license before the equipment or technology can be transferred, since all goods and services leaving the United States require some form of export licensing. An export license is essentially an authorization from the US government for US firms to export their products. Through the granting of export licenses, the US controls exports and re-exports of goods to protect national security, foreign policy, and short supply. The Bureau of Export Administration (BXA) is the star player in this. Reinsch (Reinsch,1995) describes the working of the BXA as one that promotes US national and economic security and foreign policy interests by managing and enforcing the Department's security-related trade and competitiveness programs. He says that the BXA plays a key role in challenging issues involving national security and nonproliferation, export growth, and high technology. The Bureau's continuing major challenge is combating the proliferation of weapons of mass destruction while furthering the growth of US exports which are critical to maintaining US leadership role in an increasingly competitive global economy.

The BXA implements the Export Administration Act (EAA) which provides for export controls on dual use goods and technology not only to fight proliferation, but also to pursue other national security and foreign policy . A major goal has been to simplify and update these controls. It also analyzes and protects the defense industrial and technology base, pursuant to the Defense Production Act and other laws. As the Defense Department increases its reliance on dual use high technology goods as part of its cost-cutting efforts, it ensures that the US remains competitive in those sectors and sub-sectors which is critical to its national security and economy.

The Export Administration Regulations (EAR) are issued by the Department of Commerce and Administered by The Bureau of Export Administration (BXA) to enforce the amended Export Administration Act of 1979. Under an interim rule, the Bureau of Export Administration has restructured and reorganized the Export Administration Regulations, the regulatory regime through which it imposes export and re-export and controls on those items and activities within its jurisdiction. This clarifies the language of the regulations, simplifies their application, and generally makes the export control regulatory regime more user-friendly.

While it is within the rights of developing countries to ask for and expect technology and its transfer for their own development, it should also be studied from the US perspective. Over the last decade the environment for the application export controls has changed and the US has found that this requires a fundamental reorientation of the policies designed to limit the export of dual-use goods and technologies leading to the proliferation of weapons. Until recently, the two superpowers had used the trade in international arms and technology and subsidization of weapons sales as an instrument of global competition. Looking at the world scenario today, the end of the Cold War does not mean that the supply and demand for weapons will subside but it does imply that this relatively predictable and controllable environment no longer exists (Reinicke, 1993). At this time, policy makers have to deal with a more unstable international climate, dominated by economic and technological factors and the resurgence of regional conflicts and instabilities.

For the United States, defense conversion and the fear of downsizing or right-sizing in defense employment increase the fear of global proliferation. For developing countries, regional instabilities and conflicts increase the dynamics of proliferation. Therefore it is obvious that the dynamics of proliferation in a post-Cold War world has now become more complex. They are driven by economic, social and technological factors, in addition to traditional politico-military security concerns. Any successful government must be able to deal with these multiple and sometimes conflicting policy objectives. In addition, the worldwide spread of technology and globalization of economy have also affected efforts to curb proliferation. The globalization of the defense industry and the resultant 'internationalization of availability" have already transformed the external environment of proliferation and export controls. (Muroyama, 1988). For many years, the expansion of international trade accompanied by even faster growth in international financial transactions, has been one of the primary factors behind the trend toward internationalization. The US fears that these changes will make an impact on their leadership in world affairs and that they would soon lose their edge in diverse fields.

With these considerations, decisions concerning US arms transfer policies are, by law, intended to be made only on the basis of their contribution to US foreign policy objectives and not for commercial or industrial motives. When a basic policy decision is made, policy-level officials rarely encounter individual cases.

Once a policy is established, the Office of Defense Trade Controls (ODTC) will routinely deny applications for end users whose application does not meet established policy criteria. India's circumstances allow for very few cases where applications are subject to routine approval. As a result, virtually all significant license applications become subject to interagency review which can be a 'labyrinth of control' (Delgrego,1995). The ODTC refers such cases to the full range of interested agencies within the US government. It is frequently the case that a single denial recommendation will result in a rejection of the application by the Department of State. One then realizes that the decision to grant or not grant an export license may quite often rest with individuals who may interpret the rules in quite a different way than for which it was meant. These anomalies are evident in the working world by those who go through the pains of getting their licenses granted.

The defense cooperation established between the US Department Of Defense and India's Ministry of Defense has the potential to serve as an instrument to address the export license denial issue. India is grouped with a number of nations who are non-signatories of multilateral or bilateral arms / technology transfer agreements and the US government uses this as a yard stick to limit exports. Since India feels that it qualifies for a better deal, it is left to the two governments to sit down and arrive at a mutually beneficial solution.

There are major reforms still brewing in the Bureau of Export Control and some have already been introduced. On October 6, 1995, President Clinton announced a major revision of US controls on computer exports (BXA, 1995). The new regulations eliminate or significantly ease controls on computer exports to most countries in the world, except terrorist countries. In an era in which information has emerged as an increasingly important dimension of national power, US export control policy toward trade in super computers has recently turned away from traditional restrictive strategies, favoring instead an approach that will foster the wide diffusion of this technology. However, in the security sphere, concerns will necessarily arise regarding the possibility that other nations, or even some non-state actors, might use the vast power of these computational engines to enable them to develop and secretly test a wide range of weapons (Arquilla, 1996).

The export licensing process at the US Department of Defense continues to be guided by the desire to reduce the potential for competitive disadvantage to the US industry while safeguarding national security under existing regulations. Through automation and streamlining of procedures, the US has consistently reduced the length of time it takes to process an export license. But while the West-to-West dual-use application reviews are completed in only a few days, the same for West-to-East, depending on the commodity, would take in excess of 90 days.

R. J. Augustus

9
Indian R&D Scenario

Defining New And Emerging Technology

The Indian security complex has grown into a diversified set of establishments that have grown up around two models. Abraham (Abraham, 1992) describes the first model as being based on a top-down structure that is less suited to technological innovation and development since it concentrates on licensed production and production for the civilian market. The second model is a more flexible and project-oriented system that has shown its ability to produce high technology equipment. It is the successes of this latter model with its civilian scientists, that have led to concern about the growing power of India. This growth continues and is very relevant in today's technological scene.

There has been a sea change in the global technology base since private industry worldwide has been increasing its expenditures on research and development faster than the government with this being especially true of the US. R&D investments by America's western trading partners and commercial competitors have also grown rapidly. These nations have rebuilt their depleted economies and over the last fifty years have climbed back to their former pedestals as major economic powers. The defense industry is gradually losing ground in fast-moving technology as today there are perhaps as sophisticated integrated circuits in automobiles and children's toys as in missile guidance systems (CCSTG, 1993). Advanced simulation, advanced computing, manufacturing technology and materials initiatives have made a profound impact on military technological superiority. Another aspect of the sharp increase in foreign R&D spending has been the erosion of the dominance in technology enjoyed by American commercial firms in the postwar period. America's close trading partners have also become its tough technology competitors.

For the interaction between the government and private industry in India, the message is clear. As US Admiral William A. Owens (Owens, 1996) has said, "Technology innovation is an invaluable combat multiplier, both for the near-term and for the future", so military needs must determine what aspects of technology DRDO has to pursue. Defense technology will have to incorporate the newer and better technology coming out from the commercial sector. Since all the R&D laboratories in India belong to the government and little or no R&D is being carried out by the private industries, the reliance on commercial technology available out of the country is critical for the growth of the defense technology projects in India to a vast extent. Though this reliance on foreign technology is gradually reducing by a planned dependence on the local DRDO, the dependence on foreign firms is still very much present and will take a long time to ease off.

So-called "spin-offs" from defense to commerce are fewer and less important today than in previous decades. In some fields the magnitude of defense's investments, and its willingness to sometimes seek high potential payoff in return for high financial risk, in a way that the private sector will not dare consider, can make a contribution that benefits the whole nation. As in other countries, defense technology development in India is also constrained by limited budgets even though at any point in time there are a number of technologies that are ripe for exploitation and application. Yet, endeavors must be made to explore these technologies even when budgets are limited and other opportunities cannot be pursued.

With the present thinking in India that defense technology must also support societal obligations, it is imperative that mention must be made of core technologies and dual use technology and the development and affordability of this technology. This is of utmost priority in today's defense technology scenario and is also the cause of most technology transfer problems.

Critical, Core and Dual-use Technologies

Over the years, the US has always been the pace setter in most areas of defense technology, very closely followed by Russia and most countries including the US allies vie to keep up with this pace. Developing countries tend to follow the trend in modernization of armed forces set by these countries with their particular regional circumstances being considered within the framework. The smart munitions, the digitized battlefield, changing aircraft and munitions and the explosion in information technology all based on the microchip miracle is transforming the battlefield situation awareness and battle planning (Widnall, 1995-A). Computer technology is also transforming the way militaries will plan campaigns and use capabilities to input battle objectives, targets, intelligence, forces available, munitions at hand, and the logistical constraints to rapidly determine what can and cannot be done. The range of technology utilized today by defense forces the world over represents the enormous advantage they have gained by intelligent use of commercially-developed technology. This abundance of technology has its impact on all countries that are planning their defense strategies and following the US as a technological beacon-at least for the present.

Identifying what critical technology one needs and knowing who has it is the next best thing to having that technology. Having ascertained who has what technology one wants, the next logical step in transferring technology is knowing how to acquire that technology either from national or international sources in as quick a time as possible with as minimum a fuss as possible. Analysis of the future capabilities of one's defense forces and therefore its needs and the scientific and technological opportunities that exist today leads to establishing several technology areas as requiring high priority investment.

If one attempts to look at India's technological future, one question that seems to always come up in almost every forum is : "If Indian industry is to "plug into" the international system, to get "globalized," as everyone says it should be, then should not the Indian R&D system, including the defense segment, be encouraged to plug into the international R&D system? (Santhanam, 1992)." The obvious answer is in the affirmative provided superfluous barriers and regulations are lowered or removed. The area that India should plug into should be "critical defense technologies", as it is from here that most technologies have emerged and are emerging at least from the initial concept and design stages. One can logically come to the conclusion that factors like technology status, manpower, resources, political stability, liberalized economic conditions, technology protection experience, and restraint in the export of sensitive technologies would make India a cost-effective, attractive partner for R&D cooperation in these critical technologies. If the USA is looking for off-shore R&D in critical technologies, then the Indian infrastructure and resources can be considered. This could be on agreed terms, including security aspects and could lead to products for markets in both countries and the world. Co-development leading to co-production of leading products may be the only way out in a situation extremely vulnerable to budget cuts and new competition.

While every technology can contribute to more than one area of development, there are some common technologies so versatile that their development and application can be so wide spread and embracing so as to be christened 'Core Technologies'. As far as India is concerned, core technologies are those technologies that provide the materials, electronics, software process, computing and components that are essential for meeting system needs. These technologies could also be typically dual-use in nature, serving to advance commercial products as well as military systems. Today the trend is to move in the direction of a shared national infrastructure with greater reliance on the civil sector to support defense needs. The most crucial area in defense applications, is the DRDO self-reliance program that emphasizes innovative technology development in support of improved, affordable defense capability.

Since core technology is dynamic and subject to change, one must plan for the national core technology base to be continually innovated and modified so that it does not become irrelevant or lose to external competition. One must also be aware that 'sheltered technology' as applicable to defense and strategic technologies do not contribute to many areas of development, except to the specific ones they were meant for. So even its contributions by defense sales and employment opportunities are debatable as defense manufacturing technologies are not good enough to compete with innovative and superior civilian manufacturing technologies.

Industries that once thrived on defense specific products are finding out that they cannot be sustained by current and future defense budgets. Technologies critical to defense are being developed and manufactured commercially and marketed internationally. Therefore, in the future, instead of looking for defense specific items, Indian R&D must rely as far as possible on integrating commercial products into its defense projects. There are problems from the point of view of India importing items from the US, that may have dual use applications. In its 1994 report on 'Export licensing procedures for Dual-Use Items Need to be Strengthened', the General Accounting Office defines dual use items as: 'Dual use items may consist of equipment materials and technical data that have civilian uses which can also be used for the design, fabrication, testing and production of nuclear explosives or special nuclear material' (GAO-1,1994) Under this definition, it is quite obvious that India cannot be cleared to receive quite a number of items, since the US feels that India has a large nuclear program.

So commercial off the shelf (COTS) items are now the current defense interest. COTS will give rise to both reduced cost of product and technological sophistication. In the US, emphasis is now placed on COTS technology for government use to reduce the cost of items purchased by the federal government. The US feels that a common commercial and defense industrial base will serve her defense needs better, and it will enhance US economic competitiveness. Hence, industry will have the benefit of combined, larger markets.

There is a change in the way Indian engineers too are planning their programs today. They are paying more attention to commercial aspects of technology and this is causing a change in the management of all defense technology related programs and issues. While defense R&D in India has been reemphasized by reforms, according to Eric Arnet of SIPRI (Haniffa, 1996-A,B) it still remains apparently immune to managerial improvements even as it consumes an unprecedented proportion of total R&D funding. Indian defense R&D has started the process of encouraging the commercial sector so that a common industrial base can be built. Defense R&D will benefit by utilizing commercial practices, processes, and products, and by developing, where possible, technology that can be the base for both military and commercial products.

India's March To Self-Reliance

Military needs must determine what aspects of technology DRDO needs to pursue and with what priority. It is the Indian 'Jawan' who enunciates those needs in today's combat environment of local warfare, major regional conflicts, proliferation of weapons of mass destruction and peacekeeping operations. One would find the scene easy to understand if there was enough resources and technological capability. Because of the paucity on both scores, Indian strategy has to have a long term approach while being prepared for a short term threat. So what should be India's strategy for the longer term? One known condition is that policy of fielding weapon systems technologically equivalent to that in the West will not change knowing quite well that second-best technology just will not make it in modern warfare. India's challenge will be to achieve technological capability for its forces within an affordable budget. Hence DRDO needs to revamp its system in-order to come abreast with defense technology capabilities of the western countries.

While firms in high technology industries are in many cases being forced to invest even more than they used to in product and process development in-order to stay ahead of, or keep up with the pack, companies that used to support significant basic research seem to be withdrawing from that. As Ostry and Nelson explain (Ostry & Nelson, 1995), the combination of very strong competition and less ability to prevent rivals from finding out what one is doing in research is enough to drive companies out of the basic research business. Many firms and nations in the name of competition focus their R&D on technologies that offer clear and immediate commercial payoffs or have to meet government project deadlines. This is very true of the Indian defense R&D where government financed programs that support research are basically application oriented with fixed time frames. While realizing this necessity, it must be emphasized that due consideration, planning and funding must go into basic research and long term R&D for the overall growth of a strong technology base. This also seems to be a universal problem that calls for more coordination and cooperation than governments have thus far given as they seem to be shifting their research goals toward applied and shorter runs. Further, basic and long run R&D calls for more funding to strengthen defense and hence the military. Though more funding may seem to satisfy the current need, Spinney (Spinney, 1985) generally argues that unless governments, defense departments and contractors change the way they do business, more funding could easily make the R&D problem worse.

Soon after independence, the Indian leadership decided to license produce or indigenously develop defense and other equipment to the greatest extent possible. It was felt that reliance on high-technology and defense imports could only lead to India's continued economic inequality relative to the developed nations. India has therefore focused on high-tech sectors such as atomic science, space applications, and in defense, aircraft and missiles once thought to be the domain of only the most advanced countries. The objective of self-reliance appeals to a nation that detests being in a position of dependence as a result of its colonial history. "Self-reliance" does have a distinctive ideological connotation in the Indian context, but it may also be seen as a means to achieve the objectives of economic and technological development, and the protection of Indian foreign policy from external pressures. The Indian missile programs both reflect and are inspired by the push for technological and economic self-reliance while ensuring national security.

It was also rightfully felt that any success in high-technology defense industries would have a positive impact in other areas of the economy. This would provide the country with skilled scientific manpower, a technology awareness and direct technology transfer from one sector to the next. Like the software capability available in India today, achieving and maintaining Indian specific technological capabilities generates confidence which contributes to other areas of the economy   Defense and industrial self-reliance has obvious implications for India's stand on foreign policy issues. Former Prime Minister Indira Gandhi once described Indian non-alignment as a policy not an objective.

R. J. Augustus

The Indian objective is freedom of judgment and of action to safeguard the nation's basic interest. Looked at from a slightly different perspective, non-alignment helps to inhibit the ability of developed states to derive political benefit from technology transfer and gives India some more breathing and growing space. India thus views indigenous weapons production as a means to insulate itself from political pressures which foreign suppliers or their governments could exert. The US restrictions on the use of a Cray-XMP-14 computer in the 1980 and a denial for one in 1990, the reversal of a decision to sell the Combined Acceleration Vibration Climatic test System (CAVCTS) in 1989, followed by sanctions on India after ISRO purchased cryogenic engine technology from Russia and the not infrequent denials to the Indian labs even for the most non-controversial systems only reinforce this idea.

The Indian government views such sanctions as a clear attempt to modify their own internal and external behavior, and that vulnerability compels greater self-reliance in programs integral to the defense and development of the country. The Indian Integrated Guided Missile Development Program (IGMDP) is an attempt to protect India from dependence on outside, unreliable suppliers. Brahma Chellaney (Brahma Chellaney, 1991) describes how India has emerged almost self-reliant in solid propellants and satellite technology and has made rapid strides in earth-storable liquid propulsion technology, navigation, guidance and control system, and telemetry, tracking and command systems. In barely seven years after its establishment, the ballistic missile program flight-tested five separate missile systems, India has built a latent intercontinental ballistic missile capability despite some launch-vehicle setbacks .

One final aspect of self-reliance expressed in some Indian circles is that an indigenous missile program and the high-technology it develops and incorporates increases Indian foreign policy flexibility and impact. DRDO's Scientific Adviser Dr. Kalam reasons that the advanced industrial countries are only willing to supply "second-hand" technology, and with such technology one can never hope to catch-up or have access to the state of the art equipment. Hence developing countries will always be behind. He believes world affairs are largely determined by the premise that "technology respects technology and strength respects strength" (Victor,1990).

Indian power is dependent on its own ability to make progress in high-tech sectors such as missile and space. One practical effect of the missile program is that the development of the Agni will lead to greater democratization of the international decision-making process since it would not be meaningful to have future arms control negotiations without Indian participation. Indian military officials are seeking to reduce their reliance on imports by easing military specifications that have precluded local commercial companies from providing ammunition and supplies to the country's armed forces (Raghuvanshi,1995). The military and civilian sectors could enjoy greater efficiency and economies of scale if manufacturing specifications are integrated and coordinated. In "Army Industry Partnership," the first Army-sponsored seminar aimed at cultivating civilian supply sources, service officials appealed to industry to adopt long-term strategies that will accommodate military purchasing requirements.

There is a need to decrease the dependence of the Indian armed forces on imported systems with the emphasis that technology transfer from the Defence Research and Development

Organization (DRDO) should be organized in ways that maximize benefits to local industry. There must be an all-out effort to attract commercial supply sources to dovetail with the national goal of self-reliance for defense purposes by 2005, a policy that has been endorsed repeatedly by the Indian Parliament. However this has come in for criticism by Indian and foreign agencies (Haniffa, 1996-B) even though senior US officials have on more than one occasion declared that the USA supports a 'robust' Indian defense capability (Jasjit Singh, 1996).

Army and industry officials have asked for a swift technology transfer from the government run DRDO to private industry. Military officials say the Army has earmarked $33.3 million for purchases from the commercial sector out of an annual $933.3 million acquisition budget. Indian Army-industry relations stress the need for developing traditional commercial ties and exploring possibilities of new joint ventures. Local industries are expected to become a major supplier of spare parts and equipment. It is also a matter of interest and concern for the planners that major systems that require tremendous infrastructure investments would probably dissuade commercial enterprises from becoming military suppliers. However one discordant note in this harmony is the late 1997 scheduled defense talks between India and Russia to discuss possible long term transfers of military hardware to India. These talks are part of an existing ``military-technical" forum and though considered to be necessary in some circles, might prove to be self-defeating for India in the long run.

India's R&D Strategy

India's military industry complex is one of the oldest, largest, and most diversified in the developing world. The expansion of India's defense industrial capacity, particularly in the 1980s, was largely driven by India's dominant neighbors, China and Pakistan. To secure its strategic objectives, the Indian Government has established a large scale defense industry sector that includes 9 state-owned defense industries, 33 ordnance factories, and 50 R&-D establishments and laboratories which is large even by standards of advanced countries (Sanders, 1990). The long-term goal has been to build an indigenous defense industrial base capable of supplying a wide range of advanced defense equipment.

India's policy of self-reliance in defense production has been complemented by imports of sophisticated weapon systems and related technologies from the Soviet and the success of this strategy is proven by India's advanced production capabilities in categories like aircraft and helicopters; main battle tanks and armored personnel carriers; diesel-powered submarines and frigates; ballistic missiles; electronic and communication equipment; and small arms, artillery and ammunition. However, the Indian defense industries remain dependent on foreign technology, particularly systems produced under license from the Soviet Union. Since 1985 the Indian Government has encouraged greater interaction between defense production and civil industry by promoting private sector participation.

Unlike many other defense producers among the newly industrializing countries, India has invested heavily in its defense R&D base to achieve self-sufficiency in defense production, and to reduce imports of foreign technologies. The DRDO functions as a central coordinating agency for the execution of all defense-related research. Interestingly, the bulk of research and development capability is mostly in government organizations with little or a few cases of R&D infrastructure in industries. Indian industries and the defense services feel that the technologies developed at government R&D establishments are not commercially viable and prefer to buy the manufacturing processes from industries abroad due to fear of risk and the effort needed for indigenous technology transfer. Policy makers feel that this situation has to be changed. The technology transfer from abroad in today's context, is only the fabrication process. Indian scientists must acknowledge the fact that indigenous high technology is the wealth of any nation and no nation will immediately like to transfer their technology to anyone. There would be a definite time delay in the willingness of industrialized countries to transfer such technology (Kalam, 1995-A). The only way for the Indian industries to have a competitive edge is by investing in R&D and working together with the Indian R&D organizations. Through such real-time interactions, the role of Indian industry would get upgraded to total system design from a limited role of production and assembly.

Some small countries that India does business with confirm that 20-30 per cent of their market economy is driven by India and some developed countries have their defense industry actuated by Indian arms requirement. Again it is felt that India needs to change this situation. Indian industries should design, develop and produce products and systems not only for the Indian consumer market but also try to capture an export market in specialized areas like defense technology. The R&D Strategy to reach this goal is sketched in Figure 20 and can be projected for

In Defense Of My Country

the multiple dimensions of technology that include research, development, technology transfer, technology absorption and production. Indian labs would have two routes they can follow. Route 'A' would be the standard licensed fabrication process that India has been following or else they could follow route 'B' which is the design route. Knowing how the design process works is what will pay off in the long haul. The tie up with industry for transfer and absorption of the technology should yield rich dividends for India's ambitions. Figure 21 describes the current practice and the real-time Technology Absorption Approach that India follows. There is a vision that Indian Industry must be transformed into a Design, Development and Manufacturing Agency to be at least on par with the rest of the world.

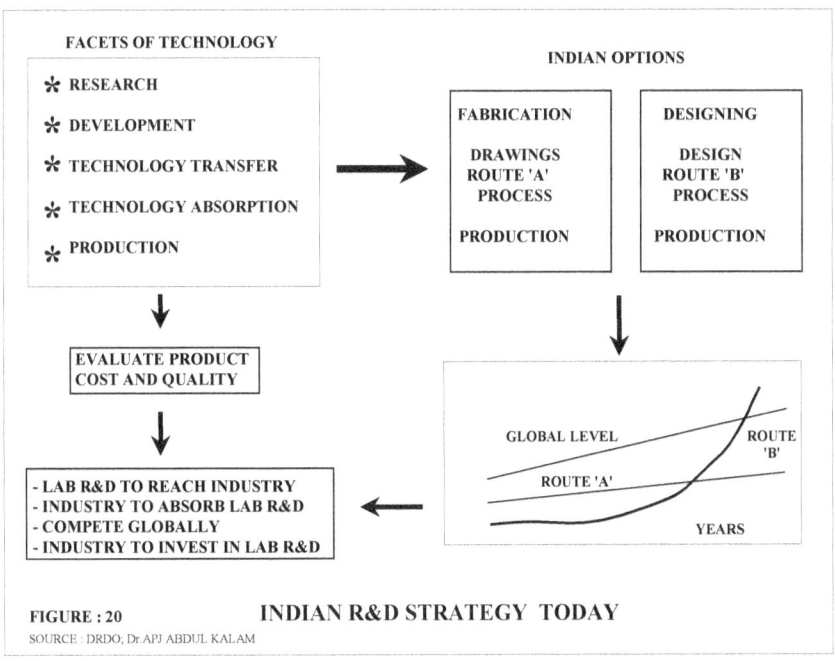

FIGURE : 20   **INDIAN R&D STRATEGY TODAY**
SOURCE : DRDO, Dr. APJ ABDUL KALAM

R. J. Augustus

Figure 22 schematically represents this DRDO vision that if followed would place Indian R&D and industry on par with the rest of the world.

History is replete with examples and even epochs of new technology generating faster economic growth and theoretical analysis generally has focused on the potential inability of private investors to appropriate fully the returns from investment in R&D. The contribution of research to a nation's economic performance depends on how well the nation's firms can utilize and commercialize research to bring out profitable new product and processes (Smith & Barfield, 1996). In the Indian context, too often due to short sightedness, R&D programs that do not lead to fielded hardware are viewed as failures. This is especially critical taking into accounts the reduced budgets and dependence on foreign sources for technology.

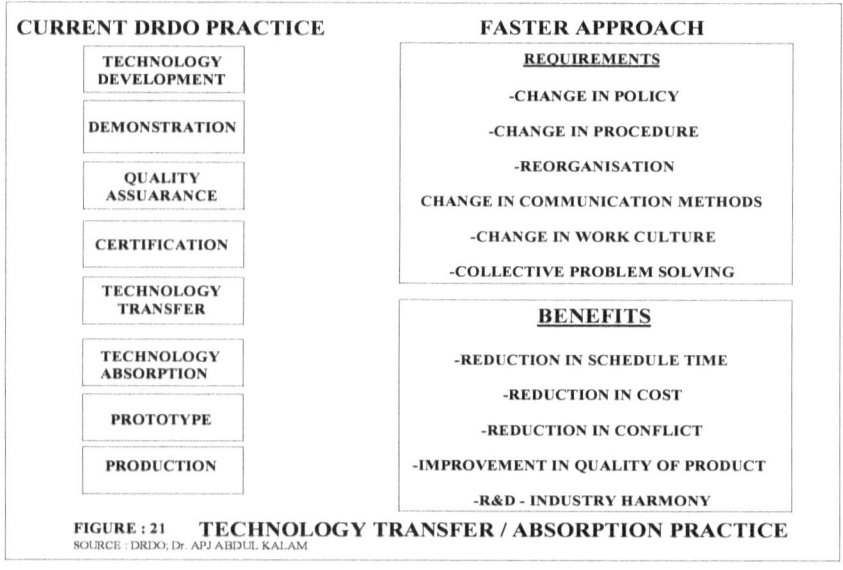

FIGURE : 21   TECHNOLOGY TRANSFER / ABSORPTION PRACTICE
SOURCE : DRDO; Dr. APJ ABDUL KALAM

In Defense Of My Country

India's Missile Thrust

Since the crux of all India's technology transfer problems as seen by Western nations, stem basically from India's thrust towards self-reliance in missile technology, it is necessary to skim through some of the salient features that motivate India to proceed in the teeth of western opposition. India's ability to indigenously design and develop sophisticated systems such as combat aircraft and missiles is the product of its long-term strategy to build technical knowledge and scientific man power within the country. Ultimately, India's ability to ensure its own security and to follow an uninfluenced foreign policy is dependent on the level the nation is self-reliant in technology and defense. As "Strength respects Strength and Technology is Strength", so a country's emphasis to achieve self-sufficiency in technology does not seem out of place.

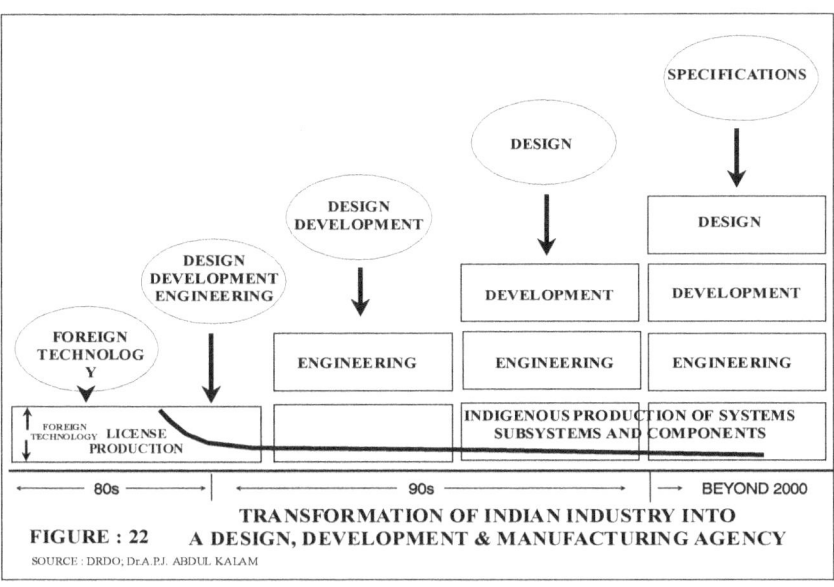

FIGURE : 22 — TRANSFORMATION OF INDIAN INDUSTRY INTO A DESIGN, DEVELOPMENT & MANUFACTURING AGENCY
SOURCE : DRDO; Dr.A.P.J. ABDUL KALAM

There are two major factors that have dictated India's establishment of its missile program. The first factor is the government's thrust for economic, scientific and technical "self-reliance." This is a national strategy that seeks to write its own political, economic and social doctrine free from any influence. The second factor is Indian military strategy and national security considerations. Like all other missile powers, India too appreciates the fact that missile capabilities enhance Indian military power which can be a strong deterrent or can turn the tide of war. As explained earlier in the chapter on India's neighborhood, Indian defense concerns have always centered around the perceived threats from China and Pakistan. Indian defense strategists are well aware of the technological revolution in modern warfare and the crucial part missiles occupy in the 21st century battle space. With so much global preoccupation with missile defenses, it would be unfair to the Indians, if they are denied the chance to protect themselves. After all history is a witness to ambitious nations looking eastward towards India. So the considerable resources, time and trained manpower devoted to the Indian IGMDP clearly demonstrate the government's disposition towards this method of defense.

According to Dr. Kalam, who the US claims was trained on missile design in the US (OTA,1993), the Gulf War has proved the cost-effectiveness of missile systems by demonstrating their ability to cause considerable damage to strategic installations. It has emphasized the Third World's strategic vulnerability vis-à-vis the United States and other militarily advanced nations. This perception reinforces the need for India to acquire even more advanced technologies and spurs India to increase her defense R&D capability. An article on 1 August in Delhi's The Economic

Times has highlighted the point that Western export control restrictions on high technology "have only helped build indigenous capability in high-tech areas. Besides all aspects of nuclear technologies, solid rocket fuel, various defense related technologies, super computing systems etc. have resulted from initiatives in response to such export restrictions."

The Right To Affordable Technology

As Kaminski says (Kaminski,1995-A) of the US, the policy of fielding technologically superior weapon systems will not change. Second-best just does not cut it in modern warfare. This sentiment is voiced in countries the world over with the Indian voice ringing loud and clear. India certainly does not want second best technology today. Every country tries to get the state of the art technology and weapon systems but at a price they can afford. So it should be clear where one is headed in the world of defense technology and equipment procurement.

India's stagnant or marginally increased budgets compel the consideration of affordability as an integral part of the technology build up and self-reliance program. The Indian defense budget, meager by world standards, for acquiring new systems and modernizing old ones must ensure that the defense services can buy more for less. In the past, threat considerations and the budget permitted the Indian R&D to focus on a program and go at it for years with all it had. Today, all programs are time and cost critical and systems must be developed at a lower cost, with a longer life span and with incremental capability through planned upgrades.

India too is in a period of transition like the rest of the world and is grappling with the political and the technological changes. Having concentrated for years on military specifications for all components and equipment, the shift in the balance between the military and commercial elements now requires the Indian national industrial base to reorganize itself and ride the global trends . Whether India is spending enough on defense research is a question that Indians are going to hear frequently in the future and it deserves to be considered seriously. While it is imperative that there should be enough provision for defense, the more important point of focus should be on what should be the defense needs of India, whether the DRDO has its sights trained on the right programs and whether it has the necessary skilled manpower structure to support these needs? As it stands today, the DRDO carries a lot of driftwood that has to be cast away before it can be molded into a corporate structure as envisaged by its policy makers.

Technology Absorption And Management

A frequently asked question by those who do business with the Defence Research and Development Organization is about the inordinate delay from the sanction to the completion of any project. Slipping schedules and cost over runs seem to be an inherent part of Indian R&D and this delay snowballs into obsolete technology. With all these problems, it seems that there is justification for the Indian defense services to look to the global market for their defense needs as they have the tremendous job of building up a state of the art defense inventory, training and defending the country.

In Defense Of My Country

Over the last few years, the top management of the DRDO have taken a close look at what ails the organization and have started to revamp the system. Analysis of past projects highlight certain problem areas. Staff requirements on a project are initially not clear and firm. It tends to change as the project progresses and the learning curve begins. DRDO tends to accommodate these changes and the associated delays that go with it. Scientists and engineers are research oriented and tend to have a poor idea of costs and time involved in large projects and are not specifically trained to deal with foreign firms on an equal footing. After all, in business one gets what one negotiates for and not what one deserves. A combination of such problems leads to disastrous cost overruns and delays. So the inherent time lag between lab and field notwithstanding, the length of time it takes for a new technology to be fielded from the DRDO to the services is very prolonged and disturbing.

Even a country like the US feels that its acquisition system is very slow (Preston, 1995) but with the changes being made, the US serves as a very good example for it has made them more efficient, improved the US business practices and allows them to buy more with less (Kaminski, 1995-A). If this is the case with the US, it is all the more reason why is it imperative to re-engineer the Indian acquisition process now. Since interaction with US defense industry on technology transfer is the prime objective, a proper environment is a prerequisite. The lead time needed to field new technology in India can be reduced by addressing three vital aspects - affordability, acquisition and absorption. Though India may plan on modernization of her defense forces, improve her R&D strategy and train her technical manpower, the seeds will never grow if they are strangled by obsolete procurement practices, strict and impractical government regulations and unnecessary paperwork. By breaking down the barriers between the defense and commercial sectors of the economy, India can make better use of the nation's resources. Indian DRDO must join hands strongly with the commercial sector as is being hesitatingly attempted in some projects currently. This joining of hands must be built into the DRDO systems so that the projects benefit from the DRDO technology base and the commercial expertise of manufacturing and marketing. If this is not done right away, there will be a division between engineers and scientists engaged in commercial and those engaged in defense work. This would prevent the kind of one to one contact that is the essence of technology transfer.

Improvements in technology now predominantly occur in the commercial sector of developed countries at a pace that Indian defense acquisition system cannot keep up with. If India is to have access to this advanced technology, India must be able to buy from commercial suppliers, who are more often than not, unwilling to change their business practices to meet limited government requirements. Indian R&D is just not a big enough market to make it worth their while. In addition even if one can figure out a way to purchase such products, the length of time taken by the process is such that the technology is often outdated by the time it is acquired. Since the defense services in India depend on the DRDO for guidance in technical issues and in the not too distant future, they will depend solely on the DRDO to meet their demands, acquisition reform must be approached in all earnestness in Indian DRDO labs as it is a very important part of the investment equation.

Indian R&D programs need to be grounded in a deep and broad-based understanding of technology and how it is evolving.

In this context, options for dramatic new military capabilities can be recognized and exploited, and India can anticipate and counter unexpected developments in the capabilities of potential adversaries. While a high technological capability remains a guiding objective, the new world demands a more balanced approach to technology, product, and process development. Lower or static budgets increase the emphasis on affordability, long lasting weapon systems, and blending of new technology into existing systems. The health of the defense industrial base also requires increased attention and this requires support to the military-civilian industrial integration by developing dual use technology, wherever possible. Close interaction with the science and engineering community outside DRDO is crucial to assure scientific progress in military-relevant fields.

All this places new demands on and requires new approaches for the management of technology resources. Looking at how developed countries are adapting to the present global scenario, there is a need for the Indian defense departments and defense agencies to also adopt some strict management guidelines. The seven management principles listed here are guide lines to drum up the best of capabilities in the short and long term, by leveraging the best resources in the nation. These action areas are:

1. Identify technology for actual requirements
2. Share information
3. work towards cost reduction of technology and fielded systems
4. Ensure technology quality
5. Strengthening the defense-commercial industrial base,
6. encourage basic and long term research projects
7. Reduce paperwork and avoid cutting 'red tape' lengthwise.

There must be a firm dedication to planning so that Indian R&D can capitalize on the new high tech modes of defense and to achieve core competencies in these areas with emphasis on information superiority. So, procurement, application and evaluation go hand in hand towards meaningful absorption of technology. Application of the technology in DRDO labs and establishments will depend to a significant degree on the technology and may necessitate an unlearning process of some fixed ideas. This would then demand a very different caliber of management. Program directors and project directors will have to be dynamic and capable of taking calculated risks - so critical in outstanding projects. DRDO should move away from a pattern of hierarchical decision making to a process where decisions are made across organizational structures by integrated product teams. It would mean a breaking down of institutional barriers and evicting parochialism . It would mean stressing lateral communication – group to group, lab to lab and organization to organization, for a house divided cannot stand. With successful implementation, this should change the way DRDO conducts business nationally and internationally.

Industrial Cooperation

Webster defines partnership as a "contract entered into by two or more persons in which each agrees to furnish a part of the capital and labor for a business enterprise and by which each shares in some fixed proportion of profits and losses." Today one can see robust partnership and cooperation amongst the armed services, between the military and other government agencies, between the military and the civil sector and even between governments because the fiscal pressures that these establishments are facing are unlikely to abate anytime soon and it is also just a smarter way to do business (Caruana, 1995).

The partnerships that may provide the greatest return on investment for both national security as well as the national economy are those burgeoning relationships between the Ministry of Defense and commercial industry. Denman (Denman, 1995) explains that in the US too, due to poor budgets and escalation, the most advanced technologies do not emerge exclusively through defense investment and defense services are no longer the dominant customer for this high technology. So DRDO needs to attract the commercial sector. DRDO is attempting this and in addition developing new partnerships with academia, with other government agencies and even with other governments. Apart from the growth of trained manpower, enduring relationships like this will be in a constant state of development and will yield results which can be translated into required defense systems.

The benefits of this relationship extend well beyond generating research results and producing scientists. Faculties at universities and their industrial counterparts play an invaluable but perhaps less visible role providing independent scientific advice to DRDO as individuals or members of advisory committees. A striking example of this is present in the Indian defense scene today with the wide range of interactions for the Light Combat Aircraft (LCA) project. More than 200 different agencies are contributing to the development of the project. It is for the first time in India that DRDO labs, Scientific and Industrial Research labs, public sector undertakings, production agencies, teaching institutions, commercial firms and foreign governments are working together. The R&D community in India finds it a long steep road from a scientific idea to a military product.

Instead of the 'can do all' syndrome, there is a change in the way defense companies now do business with the Indian defense ministry. What they need to perform need not always be the core of their business. By subcontracting these non-core capabilities they save money, share business and save their energy for the core capability. DRDO's cooperation with these commercial firms is of paramount importance as they are on the cutting edge of innovation and may not be inclined to tolerate the budgets and work culture of government defense establishments.

The armed forces and DRDO do not have a long history of working closely with industry and the methods have not always been as smooth as they should be. It is time to now start easing the process of translating scientific capability into an operational one. This streamlining is long overdue, but it has been accelerated today by a number of new changes in the Indian environment. To further accelerate this progress there should firstly be an increase in the cooperative efforts between the armed forces and DRDO. A determined effort must be made to invest in this industry and ensure that they acquire the cutting edge in those technologies that are critical to defense requirements. A government and industry partnership could very well lead to an integrated defense and commercial industrial base. It would be very much in context to emphasize that the DRDO will gradually move towards a corporate structure. This move would be in the right direction for, providing capabilities in the future. This necessitates a close relationship between the Armed Forces, DRDO and industry since technology transfer from military to civilian purposes according to Gansler (Gansler,1994-A) occurs most effectively through people working together. As a consequence, integrating staff with commercial backgrounds and those with defense productions experience has proven to be a successful strategy in the US. This is part of the modernization program that not only Indian DRDO but any other defense organization should look towards.

DRDO And Societal Goals

It was Michael Harrington, US social scientist and author who said 'If there is technical advance without social advance, there is almost automatically, an increase in human misery, in impoverishment.' The Indian defense laboratories do not limit their involvement with just defense products but also take an active part in reaching out to society at large. There are great opportunities to focus attention on social and economic concerns and less on potential military conflicts. In the next decade and those that follow, India will confront critical public policy issues that are intimately connected with advances in science and technology.

"When a nation works to relieve the pain of its people, mentally and physically, every citizen becomes rich and divine" (Kalam, 1995-B). "Missiles for Medicine" has been a highlight in Indian defense R&D where the DRDO has adapted certain missile technologies into socially useful medical products primarily to bring them within the reach of the common man. Knowledge resulting from basic research must be exploited to improve the efficiency and effectiveness with which applied research and technological development are directed to societal goals. Policy decision making will require the integration of numerous considerations, including accepted scientific knowledge, scientific uncertainty, and conflicting political, ethical, and economic values.

R. J. Augustus

Though what has occurred in the Indian defense environment is no textbook example of organizational excellence, it is not too late to realize that policy questions will not be resolved by citizens, scientists, business executives, or government officials working alone in isolation. Addressing these issues effectively will require the coordinated effort of all sectors of Indian society.

In Defense Of My Country

10
Export Controls in a New World

The subject of arms transfers is obviously of great importance to the defense industry and is always a complex and emotion laden issue. The debate on appropriate US policy involves such issues as national security, regional arms balances, the US defense industrial base, future competitiveness, jobs, ethnic politics, and morality. It is not surprising, therefore, that the debate is at times heated, and that participants in the debate are all too often talking past each other. This may not seem very strange when one recalls that at times the US has blown hot and blown cold over the issue of arms and technology transfer.

What must be repeatedly brought into focus is that Indian Defense Research and Development is not looking for the import of deadly weapons but for the import of high technology components and equipment for its own exploratory defense technology research and development programs. The small number of items required is proof enough that India is definitely doing its own research and undoubtedly with the sole purpose of building up her own technology infrastructure and the small number of components are not going to arm the Indian defense forces. The US and other industrialized countries fear that this technology will be developed by India and used to build her own defense arsenal and equip her armed forces. This is true, for India like any other nation would prefer to save the foreign exchange otherwise being spent on technology and equipment purchases from foreign industrial firms with a view to improve the effectiveness of her fighters. After all, it is the readiness of the defense personnel that matters. In this regard the US outlook is not that different as voiced by Edwin Dorn, under-secretary of defense for personnel and readiness who says that the Department of Defense is obviously about troops, armaments, and military readiness, but there is a lot more to the department than what one can see on a battlefield (Dorn, 1995) and underscored by the Secretary of the Air Force who stated that President Clinton and the Secretary of Defense Perry have put a premium on keeping the troops ready with the best training and the best equipment (Widnall, 1995 - C).

So with the above points of view being shared by the US, India and so many other countries, Indian ambition of achieving technological capability is a fair and justifiable one. No country, with the skill and technical base like India's, would like to sit back and depend on external sources. At some time, developing countries will aspire to be called developed even if it is only in the technology area. This search for self-reliance and the longing to get into the international defense technology market combined with global upheavals has created a strange environment for Indian R&D.

The industrialized countries are at a delicate phase in their post-cold war relationships for they all face the challenges of scaling back and restructuring their cold war defense industrial bases. So these governments are pursuing expanded arms exports to developing countries to support their struggling arms industries and do not appear to recognize the long-term structural decline that is under way in international arms markets. This increase or decrease is debatable for some consider that in certain regions of the world, defense budgets are increasing (Johnson, 1992). This resulting competition will likely produce increased security threats in the form of technology and weapons proliferation, increased regional insecurity, and heightened political tension in several regions of the world.

Continuing reliance on the cold war formula for security may now prove to be risky not only for the Americans and Russians, but for all countries. Their once indispensable defense industrial bases have now become economic burdens as well as an impetus for weapons sales. If the industrialized countries indulge in the transfer of arms and technology it is termed as sales and the same transaction by a third world country is termed as proliferation! Instead of converting and fundamentally restructuring their respective cold war defense industries, the US and Russian governments are attempting to shore them up, in large part by expanding arms exports. Regardless of the success or failure of arms export efforts, supplier competition poses serious long-term threats to the United States, Russia, and the rest of the world. Because international arms trade has become a buyer's market, the likelihood of proliferation of advanced and even state-of-the-art weapon technologies has increased substantially.

The decline in the demand for arms exports and excess defense industrial capacity are in part symptoms of ongoing revolutions in international security and industrial technology. Fundamental changes are occurring in the global security environment, as well as in the nature of armaments and manufacturing technologies. New strategies are required for building armed forces, managing defense industries, and stemming the proliferation of advanced technologies. To throw new light on proliferation controls, a more positive approach has been projected by analysts from the Brookings Institution who have expressed a different point of view. Since export controls are becoming increasingly ineffective, they should be reduced and in their place the US should try to build an international consensus on achieving greatly increased transparency in international trade and in national industrial activities. This could mean that reduced export controls could be replaced by increased export reporting requirements, plus intensified governmental and other monitoring aimed at exposing proliferating programs to international sanctions (Steinbruner & Nolan, 1993)

The Trouble with Arms Exports

To sell or not to sell is the question. When dealing with the US, one sees a paradox in its export control policy. The US executive branch has spent two years attempting to formulate a conventional arms transfer policy for the post-Cold War period, and the 104th Congress has attempted to rewrite the Arms Export Control Act. The most important question for both is how to keep certain weapons from aggressive or irresponsible states. At the same time, it is in the US interest to encourage friendly governments to provide for their defense needs by cooperating with Washington in developing new systems or purchasing their equipment from US-based companies (DOD, 1995).

Since almost all state of the art weapons systems and technology are available from non-US sources, a meaningful and effective arms controls will only be feasible through a multilateral effort. This multilateral effort must firstly increase the transparency of trade in technology and weapons thereby coercing suppliers and purchasers to justify and limit such trade. Secondly it is imperative that all suppliers agree on what should and should not be sold. The third is to encourage regional groups of countries to boycott certain categories of weapons in their regions.

Whatever the short-term benefits of foreign arms sales, the effects of arms export competition are likely to aggravate global tension generally. Because of the overcapacity and large industrial bases in major countries, the competition among suppliers will be very destructive, regardless of the success of various countries in their export efforts. It is again stressed that in a buyer's market with intense competition among producer states, there is an increasing possibility that current weapon systems will be sold indiscriminately. One will be right in assuming that to the extent that the governments pursue arms exports instead of defense industry downsizing, they would be merely putting off inevitable costly and difficult choices that will only get more difficult and costly.

Janne Nolan (Nolan,1994) projects that even co-production and licensing arrangements for foreign production of weapons poses great proliferation risks as technologies and production techniques are also being transferred with the weapons. Since such sales would increase the global technical knowledge base, supplier countries will have to debate well before selling. For the sake of financial gains, arms could be transferred to one's potential enemies and also to regions where historic animosity exists, thereby starting an arms race that may have a very unappealing end.

Another point to be noted is that defense industries who have for years made their fortunes by trading arms, may not be inclined to relinquish the market or convert. Such industries could also act as a drag on the national budgets and economies by extracting subsidies or other support from the government. To alleviate this pressure, governments may look to developing countries to purchase a wide variety of items ranging from components to weaponry.

'International Competitiveness' has always been stressed to maintain US technical preeminence in a period of increased competition and consolidation (AIA,1994). To understand this requirement it would be useful to study just how important defense exports are to the US defense industry, and what trends might occur over the next decade. The technological revolution is affecting the nature of particular armaments, the manufacturing processes used to produce them, and the ways that industry is organized generally leading to a sea change in the nature of arms and warfare. Defense industries could simply be one part of a larger industrial base and the government may be just another customer. The larger industrial base would then take care of all government projects and R&D requirements (Gansler, 1994-B). The US with its mammoth defense budget, has always had the luxury of designing products strictly for its own defense services. With the reduced budgets and drop in domestic requirements, this may change and over the next few years, there could be a tendency for increased defense technology and equipment exports.

If one reads the tea leaves right, there is a growing awareness in industrialized countries that exports can keep down the cost of their own defense research and development, testing and defense procurements. So again one may witness actual support from the government to the defense industries in their sales efforts. The same is true of some in the US Congress, who are beginning to see lowered defense budgets result in plant layoffs and decreased work for subcontractors. Exports to friendly countries can help keep those production lines rolling. So there could be a possibility of 'business as usual' provided there is some justification for an export against a reasonable requirement.

But the security and industrial-technological revolutions will require radical transformations to the manner in which the US and Indian defense planning systems provide for their security. Although the particular circumstances in the two states differ, both will have to rethink their new relationship. A new approach to the defense industrial base will need to be adopted that stresses integration of the defense and civil-commercial sectors. The security threats that US feels it will face are more likely to be lower in intensity and stem from problems of technology proliferation or the political disintegration of states and analysts feel that "these problems are related more to conditions such as endemic economic austerity in highly populated regions than traditional state-versus-state conflict"(Carter, 1992 ; Steinbruner, 1994).

The Third World needs to assess its collective and individual strengths and weaknesses against this back drop and institute their policies to address themselves to future developments with care. The world is changing. That it changes to the advantage of the developing countries would depend on how well they manage their resources and create conditions to withstand efforts to their further exploitation. The era of global cooperation, howsoever desirable, is obviously on the anvil. The 21st century appears to be heading for intensifying conflict and only the causes and means of the conflict will differ (Nayyar,1995). In this context it is felt that most of the recent proposals for arms control in Asia would diminish Washington's ability to respond to threats in Asia (Fisher, 1991).

The overall benefits to national security of applying export controls come at a price to the companies and industries whose products are controlled. The difficult task for both US Congress and the executive branch is to design an export control system that serves US security interests but also takes due account of economic interests and fairness to regulated exporters. The task is made more difficult by the inherent problems in trying to estimate both the benefits and the costs of export controls.

US Arms Sales

When one considers US arms sales, there are certain facts that have to be considered in-order to get the right picture. Firstly, there is a belief that the US is the world's leading arms seller and must show greater self-restraint if it is to convince others to do likewise. The US over the past decade or so has accounted for between 20-25% of world trade in arms. The Soviets accounted for between 40-50% according to Johnson (Johnson, 1992). It is only since the breakup of the Soviet Union that US sales have shown an increase. This is so because most sales data today will list the former Soviet states separately. Hence the US feels that it has been among the most restrictive of nations in its arms transfer polices, and other countries, rather than following its lead, have jumped into markets thereby denying US producers valuable business. This is evident in the way developing nations, India included, look towards purchases from Western Europe when any item is denied to them by the US. Americans have to also consider the defense-related employment in the United States which is declining at a rate of twenty thousand jobs per month. Such employment has fallen 20 percent since 1987, including an 11 percent drop in 1993 alone (Lubman, 1994). All these are good motives as any for pursuing sales. One good example is the US Air Force hope of using revenues from sales of its used redundant aircraft to finance its future development projects (Lewthwaite, 1994).

However, taking the above points into consideration, no one in the US industry believes that exports can offset the decline in domestic sales. Domestic procurement is likely to fall by 30-40 billion from 1985 levels. Exports are likely to remain steady or at best increase by a billion or two. But those exports tend to include major platforms which are already in production, and hence keep lines open and skilled blue collar workers employed. Statistics over the last few years show that in 1994 the US made 56 % of worldwide arms sales (Hartman,1996). Worldwide weapons transfers fell from $28.4 billion in 1993 to $22 billion in 1994 (ACDA, 1995) and US arms transfers worth $12.4 billion accounted for the 56 % mentioned above.

US industry does not believe that increasing its export promotion efforts will have much impact on the total size of the world export market. Rather its efforts are aimed at encouraging countries to look favorably on US products when it comes to using the defense funds they do spend, rather than those of its competitors. Though at times even the US 'ducks the law' and plans technology sales (Rajghatta, 1996), it should also be noted that the US Defense department is so flush with weapons that it has been giving them away free (McCarthy, 1996). The Arms sales Monitoring Project of the Federation of American Scientists found that in the past six years $7.0 billion worth of military hardware

R. J. Augustus

ranging from planes to pistols was transferred abroad, mostly to developing nations that either paid nothing or won supersaver discounts. Figure 23 represents the foreign military sales agreements between the United States and other countries while

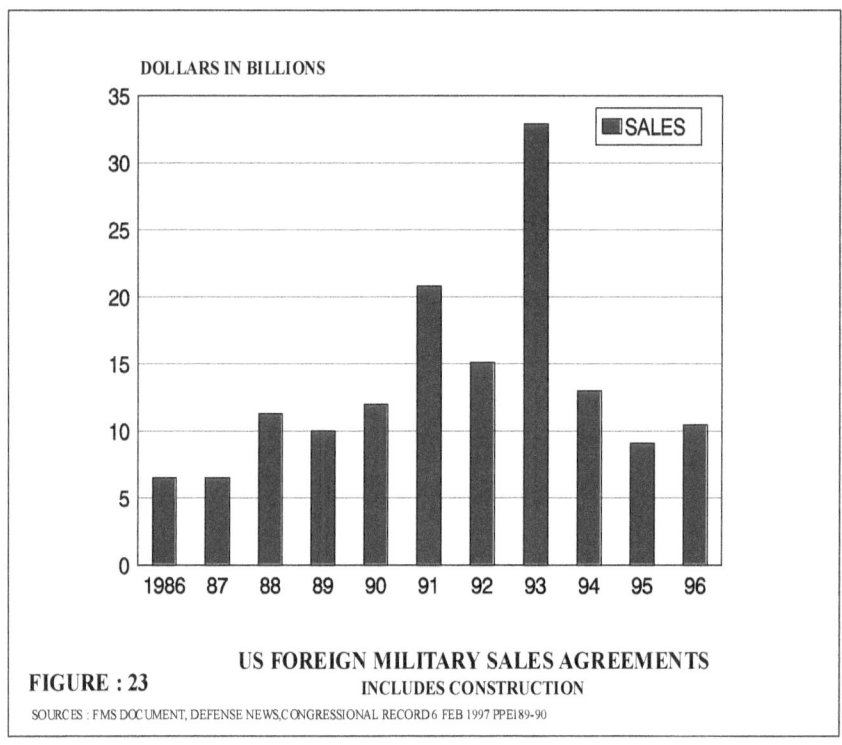

FIGURE : 23 — US FOREIGN MILITARY SALES AGREEMENTS
INCLUDES CONSTRUCTION
SOURCES : FMS DOCUMENT, DEFENSE NEWS,CONGRESSIONAL RECORD 6 FEB 1997 PPE189-90

Figure 24 represents the actual sales deliveries for the period 1986 to 1995 as published by the Department of Defense. Yet most Americans would be shocked to discover that 85 percent of US weapons are sold to countries the State Department deems undemocratic and also sold routinely to countries repressing their own citizens without any or a minimum of a congressional debate (Washburn, 1996).

In Defense Of My Country

The US has its own criteria when arms exports are involved. The fact sheet issued by the Office of the Press Secretary of the White House on 17 February, 1995 specifies that each case of arms transfers will be reviewed on a case by case basis and that it will be guided by certain criteria. Some of the highlights of this criteria specify that the transfer

1. must be consistency with international agreements and arms control initiatives,
2. must meet US and recipient country's security requirements,
3. must not imperil

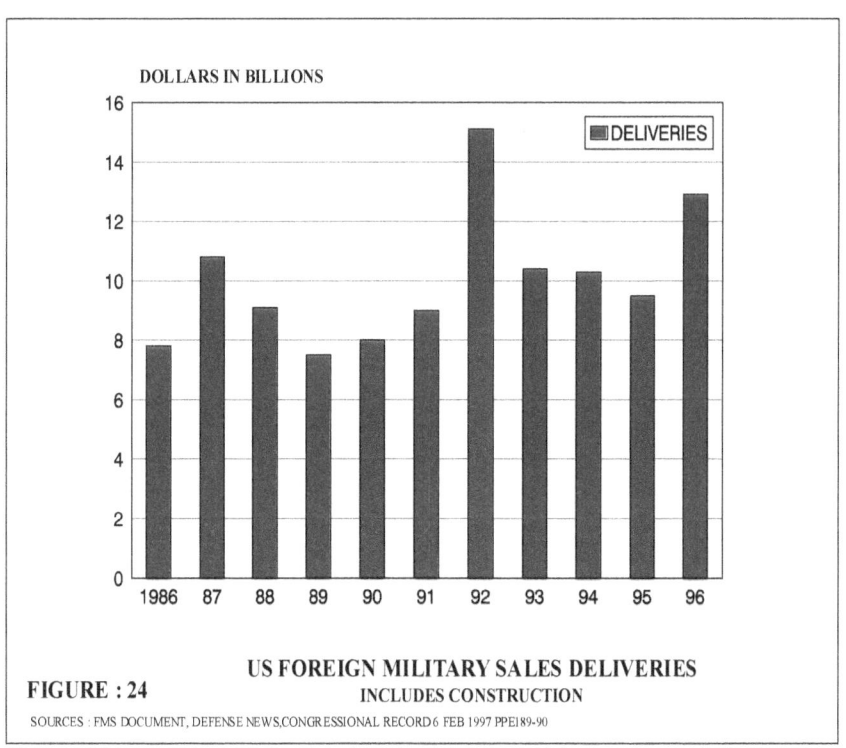

FIGURE : 24 — US FOREIGN MILITARY SALES DELIVERIES
INCLUDES CONSTRUCTION
SOURCES : FMS DOCUMENT, DEFENSE NEWS, CONGRESSIONAL RECORD 6 FEB 1997 PPE189-90

regional stability, lead to an arms race or cause tension, 4. must meet US foreign policy requirements, 5. must not compromise US technological advantage and capabilities and must not undermine US industry, 6. transferred technology must be protected as per US requirements and also restricts third party transfer is intended, 7. transfer must not reveal US system vulnerabilities thereby effecting US operational capabilities in the event of compromise, 8. transfer must not negatively impact the recipient taking into account the human rights, terrorism and proliferation record of the recipient, 9. must consider the availability of comparable systems from foreign suppliers and 10. recipient must be capable of putting the system to use for its specific purpose.

The US criteria is also very specific regarding the technology upgrades required by other countries. This has specific interest to the Indian defense scene that required an upgrade for its MiG fleet of aircraft. In addition to the above general criteria, the White House fact sheet has put forth the guidelines for upgrades. For any upgrade consideration, the scope of the total program must be clearly defined and should be consistent with general conventional arms transfer criteria outlined above. If the requested upgrade leads to a system which has a greater capability than a system the US is willing to sell, then one can expect a denial and this holds good even for upgrades by foreign contractors. All this is to avoid the technology transfer that automatically accompanies an upgrade and needless to say, any violations will automatically invite sanctions.

US Industry-Government Interaction

In discussions with various representatives of US defense firms that the DRDO does business with, it is understood that the US industry assumes that the US government will continue to pursue an active policy of encouraging sensible arms limitation agreements among supplier and user groups. But these industry representatives feel that the US government should continue to allow US defense equipment to be made available to friendly countries which have legitimate defense requirements. In such instances industry would argue that it is appropriate for the US government to work with its industry to see that such friendly countries purchase US products rather than those of its competitors.

Johnson (Johnson,1992) advocates the export of US defense equipment to friendly countries so that they can increase their ability to defend themselves and make joint activities easier through standardization of equipment. He also projects that the US industry feels that such sales could, apart from the economic benefit, increase US influence on the purchasing country's actions, both because of the ties that are established between the US defense establishment and that of the other country, and also because the purchaser's equipment will not be operable long without continued US support. Though it has been voiced with all good intent to maintain the lead by US industry, it is precisely for these reasons that developing countries are looking for second sourcing and also going in for indigenous design, development and production since no country would like to be held hostage in this manner.

However, seen from the US government perspective, export controls on dual-use goods, technology, and software will continue to be an extremely useful tool in US efforts to stem the proliferation of weapons of mass destruction and missiles that can deliver them. OTA's report on The Proliferation of Weapons of Mass Destruction (OTA, 1993) bears this out. If export controls are going to be effective against any proliferation, it will have to be tailored for a wide variety of factors. The US will have to consider the capabilities of the country, the scope of the country's programs and the determination behind this, whether the technology can be controlled, whether it is worth controlling or is it for cosmetics and whether the US can persuade other countries to go along with her and above all export controls must carry the weight of complete conviction. If after all deductions, export controls are still applied, then there could be different reactions from different countries. In some cases the proliferation may stop. In others it may be camouflaged. In some cases it could cause a delay to the country's program which can be used to persuade the country by other channels to stop it programs. But if any country sincerely believes that its programs are absolutely necessary for its wellbeing, then coercive export controls is a nonstarter.

In the US, the Code of Conduct (FAS,1995) is supposed to prohibit arms exports to any government that does not meet the criteria set out in the law, unless the President exempts a country and Congress passes a law affirming that exemption. The four conditions a country must meet in-order to be eligible for US weapons are: 1. democratic form of government, 2. respect for basic human rights of citizens, 3. non-aggression (against other states) and 4. full participation in the UN Register of Conventional Arms. The Code's criteria are all primary foreign policy tenets of past and present US administrations. Nevertheless, as mentioned earlier 85% of US arms transfers during 1990-95 went to states which did not meet the Code's criteria.

The Pentagon believes that India, in spite of having the capability and the urge to build nuclear weapons and ballistic missiles, will neither test nor deploy these weapons of mass destruction any time in the near future (Haniffa, 1996-D). In a report titled "Proliferation threat and Response" (DOD, 1996) it is mentioned that New Delhi's willingness to refrain from conducting additional nuclear tests could inhibit the development of such weapons. The report also mentions that India has one of the more self-sufficient domestic missile production programs in the developing world and New Delhi will continue to be largely unaffected by multilateral control regimes and denying access to related technology will delay, but not stop efforts to improve missiles now in development.

Commercial Or Foreign Military Sales

So then, if one is looking for procurement of any type of equipment or technology from the US, it can be done through the Foreign Military Sales (FMS) or through a Direct Commercial Sales (DCS). There seems to be a sort of belief among third world countries that any type of technology will be available for transfer from the US provided they go through a FMS rather than a DCS. Though this has not been substantiated, perhaps a look at the FMS will serve to erase some of the misconceptions.

Unless an item or service is available only via the FMS, there are no binding rules for a foreign government to do so and invariably this will not be the case. The choice of FMS or direct commercial procurement channels by a foreign government depends to a large degree on whether US defense consultancy will be needed during the procurement planning phase. The selection of the FMS or DCS for a particular procurement does not necessarily mean the foreign government has to follow the same path every time, even though once an FMS case is open, it can be used for other transaction resulting in less paper work. Each purchase and the method of purchase will have a set of pros and cons which will have to be evaluated by the government intending the purchase. The final decision on procurement channels tends to vary from country to country, and even from purchase to purchase (DISAM,1993). Early 1998 has brought grief to the Foreign Military Sales clan. With the cry from several countries that the overheads and 'administrative costs' are too high, international customers ultimately will determine whether or not the Pentagon remains in the arms sales business over the next few years. But whether these customers go through the FMS or the DCS, they will have to go through the arms control procedure before an export license is issued.

India like many other countries has chosen to go the FMS way in some of its major projects. Generally, even though the FMS path is quite tedious, responsibilities are clearly defined and above all one is dealing directly with the US government. From the foreign purchaser's perspective, the most important of these considerations are summarized below. Figure 25 represents a flow chart of the FMS process using India as the purchasing government. Under the FMS system, purchases for foreign governments are made by a well-established US DOD contracting network. There is no involvement of the foreign government in the local contract negotiations and their responsibility is limited to agreeing to requirements and estimated costs as they are stated in USG Letters of Offer and Acceptance (LOA). The US Department of Defense is responsible for the procurement of defense articles through the FMS system under the same system of procurement that it uses for its own procurement. However the purchaser will have to bear the administrative costs incurred by the DOD for work carried out.

Since the FMS system is based on a competitive procurement system, it is designed to assure maximum quantity purchases at the lowest feasible price. What the purchaser has to realize is that the purchase of US defense articles involves the nature of the follow-on support and training which will be required from US sources. If the system or items being purchased are being used by the US defense forces and are known to require substantial logistical, technical, and training support, an FMS purchase might prove to be the logical choice for it would give the purchaser the opportunity to make use of the US experience and facilities.

The purchaser is at liberty to call in any commercial company in the US they want to consider to get a good deal and this need not be the exact way DOD goes about its business. However, the greater the experience and skill level of the purchaser's negotiating staff in the Embassy or a visiting team, the greater will be the level of industry participation and competition. This can result in a very good purchase price. Under a Cooperative Logistics Supply Support Arrangement (CLSSA), the entire DOD inventory and contracting system can be drawn upon in support of the purchaser's requirements, and this can be accomplished simply by the submission of requisitions for individual parts. In effect, the DOD logistics structure serves as a procurement staff for the purchaser by procuring his required individual items from the current US sources.

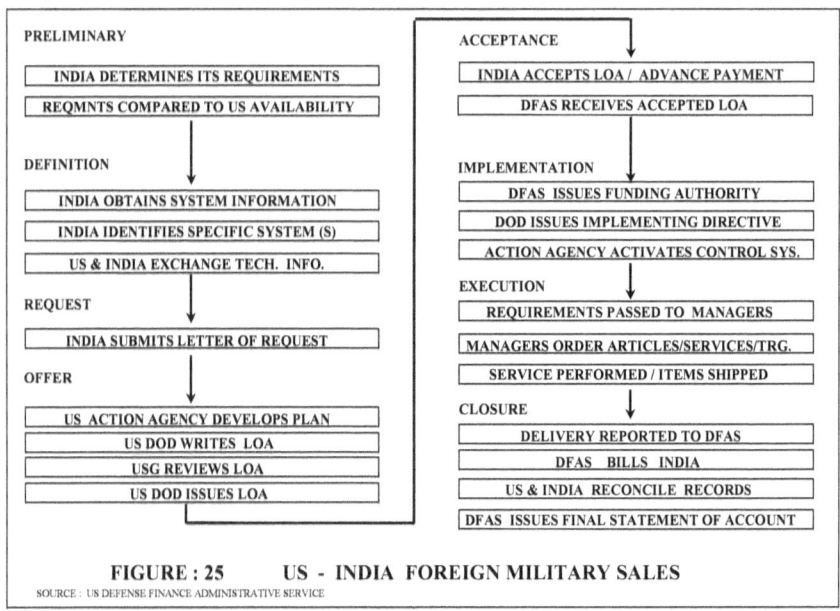

FIGURE : 25   US - INDIA FOREIGN MILITARY SALES

As far as the cost is concerned, it is difficult to predict for any particular acquisition whether it would be cost effective for the customer to follow the FMS or DCS path. This is especially true in those cases where the equipment and associated services are not fully equivalent. The purchaser may be able to get a different price from the DCS depending on the system, the risks taken by the contractor and the competition

11

Indian Participation in Controls

Based on international recognition that prevention of the proliferation of weapons of mass destruction and missiles is essential to global peace, developed countries, in addition to implementing export controls based on international agreements, are also undertaking measures on their own to reinforce security export controls even for their own security. The expansion of export controls to a wide range of items, together with industrial development, has brought certain countries such as those in the former Soviet Union, Eastern Europe and Asia, into the category of supplier countries of controlled commodities. This therefore necessitates establishing security export control systems in these countries in-order to implement internationally effective security export controls on a cooperative basis. India, as explained earlier, may not be a signatory to some of the regimes, but she does strictly adhere to self-imposed guidelines in import and export controls for the good of the regional and global security.

Indian Import Controls

On November 29 1984, India and USA signed a Memorandum of Understanding (MOU) on Technology Transfer to enhance trade and cooperation in advanced technologies. The

MOU established a framework for long term Indo-US interaction in advanced technologies. It facilitates India's access to such technologies and sets out guidelines for the import of items from USA where their export is controlled.

The Implementation Procedures of the MOU oblige the Indian importer of controlled US items to give some assurances. From 1 April 1988, these assurances are communicated to the US Government through the US exporter on an Import Certificate (IC) - a document developed for MOU transactions. A specimen IC is given in Figure 26. These certificates are issued by designated Import Certificate Issuing Authorities (ICIA) in India. An application for an Import Certificate must be made in the format given in Appendix 1. The Import Certificate states that the Indian importer of controlled US items has given the following assurances:

1. to import the item into India and not to redirect it or any part of it, to another destination before its arrival in India;

2. to provide, if asked, verification that possession of the item was taken;

3. not to re-export the item without the written approval of the ICIA;

4. to obtain permission in writing from the ICIA prior to any change in end-user. The new end user shall be obliged to notify the ICIA that it agrees to the conditions contained in the IC. In some selected cases the US Government may require extra assurances beyond those mentioned in paragraph 4. and such assurances are taken directly from the Government of India.

When such assurances are required, the concerned ICIA shall require the importer to fill in details given in the indicative format at Appendix 2. In the case of private sector importers, the Import

License will be recalled and amended to include the extra assurances given directly by the Government of India and the importer will be expected to comply strictly with all the assurances given by the Government of India to the US Government.

Appendix 3 describes the steps that are involved in the issue and transmission of the Import Certificate when the Importer is the Government or a Government-controlled entity. Appendix 4 describes similar steps when the importer is a private sector entity. For components whose import from USA is controlled, the same IC procedures are applicable. Certain guidelines are also applicable for transfer within India or export of end-products containing US controlled components. In certain cases, the US Government may seek a Delivery Verification of the items imported. In such cases the importer will be asked by the US exporter to provide Delivery Verification. It is required that the Indian importer should follow the guidelines in Appendix 5 for Delivery Verification Procedures.

Export Controls in India

It must be emphasized that India does not export missiles or technology leading to the development of weapons of mass destruction, nor has it ever been a major exporter of defense equipment in the same league as the developed nations. India's well-developed missile and space infrastructure now provides an opportunity to enter this potentially lucrative international market and the Indian government could sell complete systems technologies and expertise, or technologies now license-produced by Indian private industry.

No............................. /19......

GOVERNMENT OF INDIA
IMPORT CERTIFICATE

Exporter (Name & Address)
....

Importer (Name & Address)
....

Description of Goods　　　**Quantity**　　　　　　**Value**
....

It is certified that the importer named above has undertaken:

-To Import the item into India and not to redirect it or any part of it, to another destination before its arrival in India.
-To provide, if asked, verification that possession of the item was taken.
-Not to re-export the item without the written approval of the Certificate Issuing Authority.
-Not to retransfer within India the item(s) specified in this Certificate without the written approval of the Certificate Issuing, Authority.
-To obtain permission in- writing from the Certificate Issuing Authority prior to any
change in end-user which shall be preceded by the new end-user notifying the Certificate Issuing Authority that he/she agrees to the conditions contained in this document.

Date　　Official Seal　　Signature...................................

　　　　　　　　　　　　　　Designation ……………………..

FIGURE : 26　　INDIAN IMPORT CERTIFICATE
SOURCE : DRDO, GOVERNMENT OF INDIA

In Defense Of My Country

Unintended "leakage" of privately-produced dual-use and missile technologies through the Indian export control system is a further concern. If India wants to indulge in the export of missiles and other deadly weaponry, the primary incentive would be financial gain. The lack of foreign exchange poses severe problems for the Indian economy as a whole. Profits from such exports could transfuse badly needed hard currency into the economy to pay for critical technology imports or to fund various defense programs. So factors like financial incentives, market conditions, Indian proliferation attitudes, and the structure and viability of the Indian export control system will dictate exports.

Most other countries in India's shoes would have sought to rapidly exploit serious defense exports as a means of shoring up bad foreign exchange problems. Instead, India has sold patrol boats and helicopter trainers to Nepal and Bangladesh and reportedly supplied small arms and ammunition to Malaysia, Nigeria, Jordan, Lebanon and Oman. India is also offering a fairly wide range of defense items such as engineering equipment, electronic and electrical products, general stores, naval vehicles and services, light aero vehicles and systems and other DRDO products (GOI-DDPS, 1994). Mama (Mama,1992) also mentions advance composites, such as carbon-carbon fiber which is being produced now in India. In fact there are several private industries supplying materials and technologies to the DRDO that could export their products. India perhaps could take its cue from the US and its allies who are exporting the same things.

If R&D has to sustain itself and expenditure and export targets are to be met, India must look to sell high value-added items. The Main Battle Tank, Advanced Light Helicopter and Light Combat Aircraft all fall into this category, but are still under development and thus not yet ready for export. Export of non-ballistic systems or technologies is also an option. If it was really so inclined, India can earn foreign exchange by trying to export the Trishul and Akash surface to air missiles being developed under the IGMDP. The Pilotless Target Aircraft (PTA) 'Lakshya' could also be a prime candidate for export (WWR, 1990).

Technologies used in launch vehicle production are part of the legitimate civilian infrastructure, yet they have "dual-use" applicability for military purposes and western countries feel that the nature of the Indian space production infrastructure could contribute to a proliferation problem, especially since the commercial sector is involved. As a matter of Indian Space Research Organization (ISRO) policy, hardware fabrication for all Indian space programs is carried out, to the greatest extent possible, through subcontracting to Indian industry (UNISPACE, 1981). Through technical consultancy, technology transfer, and long-term buy back programs, ISRO has introduced hundreds of launch vehicle and process technologies to public and private firms. A major objective of this ISRO-industry partnership is to promote and tap export markets for space goods and services.

The Indian missile and space programs have benefited from significant capital investment but have yet to recover any appreciable external returns. The market for Indian missiles or missile technology might include any one of several countries and these programs can immediately offer quality items at global competitive prices. Developing countries that wish to advance their space launch capabilities may be potential buyers. India however does have a strict export control procedure in place with controls being applied by the ministry of commerce (GOI-MCOM,1995). Of course, if there is any thinking in this direction, the new government of India would have to seriously weigh the profound international repercussions that are linked to the export of missile systems or technologies.

R. J. Augustus

12
Strengthening Indo-US Relations

Any relationship requires a build of confidence till a stability sets in over a period of time. So too in the relationship between two countries. Confidence Building Measures (CBMs) have played an essential role in improving relations between the East and West and today they are emerging as a preeminent means of nations working together (Krepon,1993). It is also assumed that this concept of confidence building implies that countries that have differences of opinions today could cooperate with each other at some time. Fisher states that by requiring both countries to cooperate on military matters, CBMs could embody and project notions of shared interest - a concept of common security (Fisher C.S.,1993). However the negotiation and implementation of confidence building measures must be approached with all sincerity and political determination.

Today, there is peace, a decline in insurgencies and an increase in bilateral and multilateral cooperation after the Cold War ideological battle. Historical enemies are now trading partners with all parts of the hemisphere reaching out to one another in a growing harmony of interests. So this seems to be a time of reconciliation, renaissance and new relationships built on trust, cooperation and consensus. The defense and military establishments of the world's nations have started to play a critical role in advancing this concert. Secretary of Defense, Perry (Perry, 1995-B) says that as defense ministries and armed forces protect the security of nations in this new era, they can also work together to advance common security interests and to strengthen security relationship.

Transparency and confidence-building measures, defense cooperation and reorganization of defense forces are the three areas in which elaborate work is required. Transparency is an unusual concept when it comes to defense as the art of war involves secrecy and surprise. Transparency avoids surprises to others and nations will not arm and act for fear of the unknown. Laurance in his essay on 'Transparency in Armaments' has clearly put forth certain basic strategies for enhancing transparency (Laurance, 1992). This, if followed by all nations, would leave nothing ambiguous and would lead to better understanding when one talks of technology transfer among nations. In this context, Americans feel that they alone make available large amounts of information about exports of defense equipment and have strongly supported an effort to provide the world more information through a UN register of conventional arms (Chalmers, 1993; IDR,1994). The US defense budget is described as an open document which is reassuring to neighboring countries, because they can better assess their own security situation when they know the size and capability of the security forces in the region. One of the factors which drove the arms race during the Cold War was wrong estimates of the capabilities and the intentions of other countries, because of the secrecy in which each of the arms budgets were shrouded. In the face of secrecy, military leaders imagined the worst case scenario and tried to build up their forces and equipment, thus starting an arms race. Transparency avoids this and perhaps it can also act as a deterrent.

There are frequent security challenges, around the world and the second area of defense cooperation addresses these challenges by nations working together. It is more efficient this way and it builds trust and confidence among participating nations. One of the best examples of defense cooperation is peacekeeping operations around the world.

The third area is armed forces in the 21st century democracies. To varying degrees, defense and military establishments are facing dramatic changes as defense forces are reduced and reconfigured for the missions that nations try to foresee in the next century based on present regional and international relations. It would be very difficult to abolish military forces when one sees the worlds track record. But missions must be redefined based on scarce resources keeping in mind the right of every country to determine its own security.

Indo-US Defense Cooperation

Military to military contacts between the United States and India go back even prior to India's independence in 1947 because a very large force of United States troops was garrisoned in India during the Second World War to participate in operations in the China-Burma-India theater. After 1947, a normal contact by exchange of students continued between the United States military and the Indian Defense Services. Considerable logistic support was offered to India during the 1962 Sino-Indian conflict. Thereafter, relations continued at the defense level despite difference on policies during the Cold War. It may not be wishful thinking to ask that the relationship between India and the United States should follow the trend it took during the Truman and the Kennedy presidencies which looked for a more active policy towards India, for the present US administration seems to be on the brink of similarly reaching out to India again.

While there have been ups and downs in Indo-US defense relations over the years, the 1980s proved to be some sort of a turning point in the relations, for, defense cooperation underwent a noticeable change during the era of "leadership democracy" when Indian prime Minister Mrs. Gandhi's rapport with President Reagan resulted in the signing of the Memorandum of Understanding (MOU) on technology transfer in 1984 and the Science and Technology Initiative (STI), both landmarks in Indo-US technical cooperation. The visit of Talbot Lindstrom of the Defense Technology Security Administration and his report on technology cooperation followed by the visit of Prime Minister Rajiv Gandhi to United States in October 1987 initiated the Mission Areas approach to the defense technology cooperation. The two countries adopted the 'Mission Area Cooperation' approach and the interaction under these Mission Areas were specifically identified to keep the technology cooperation well in focus. The three Mission areas such as aircraft technology, third generation anti-tank systems and Test and Evaluation were identified. While there has been a substantial amount of cooperation and technology transfer on the Indian Light Combat Aircraft (LCA) project under the first area, nothing much was done until recently some interaction was started in the third area.

The US priority of defense cooperation with India was plainly in the field of service to service cooperation and was rather reluctant to consider the transfer of technology and arms supply.

Former Joint Chiefs of Staff General Colin Powell summed up these thoughts when he told the Indian Ambassador S.S. Ray in December 1992 that "a strong and secure bridge must be built before too many vehicles are put on it." A dramatic change in Indo US Defense Cooperation came about with the visit of the Indian Prime Minister Mr. P.V. Narasimha Rao to the United States in May 1994. This was a historic visit because there was a symmetry of views expressed by both the Indian Prime Minister and President Clinton in their joint statement issued on May 19, 1994 and the Secretary of Defense William J. Perry suggested that a closer Indo-US Defense Cooperation was desirable. This resulted in a visit by Dr. William J. Perry to India in January 1995 and the signing of the first ever Agreed Minute on Indo-US Defense Cooperation. These Minutes visualized cooperation between the Indian Ministry of Defense (MOD) and the US Department of Defense (DOD), defense services to defense services and in the area of joint research, development and production. Admiral Macke, CINC, Pacific Command, felt that the earlier series of military exercises and contact would expand under the 'Agreed Minutes' which was, a blueprint for enhanced service-to-service, civilian-to-civilian and defense production and research cooperation (Macke, 1995).

With the cold war over, the security relationship between India and the United States may be considered as a fairly good example of how military contacts can promote and enhance a broader political engagement. The earlier series of military exercises and contact have expanded under the 'Agreed Minute' and as a result of this initiative, Joint Steering Committees were set up for the Army, Air Force and Navy to coordinate Indo-US Defense Cooperation in the fields of personnel exchanges, joint training and exercises and information exchange. Another area of mutual interest has been cooperation in peace keeping where India has a long and distinguished record of supporting UN Peace Keeping.

Following the signing of the Minutes, an Indo-US Defense Policy Group (DPG) and a Joint Technical Group (JTG) were established. The DPG is responsible for all policy decisions while the JTG has the responsibility for all the technology interactions. The JTG operates under guidelines approved by the Indo-US Defense Policy Group (DPG). The goals of the JTG are to: 1. Establish a framework for bilateral cooperation in defense technologies; 2. Develop activities that will lead to substantive cooperation and monitor progress periodically and 3. To provide a forum in which the policies, plans and requirements of both sides can be discussed.

The first meeting of the DPG and the JTG was held in mid-September 1995, to discuss security policy and other strategic issues of common interest. Both countries agreed that defense cooperation was important and worth pursuing despite recognized differences of perception on nuclear and missile issues. Since 1995, there has been a thrust in another Mission Area which was considered not too controversial. This was on Test and Evaluation of defense systems. Successive JTG meetings have been in held in Oct 1996 and January 1998.

While the interactions have been proceeding 'smoothly' taking into consideration the two bureaucratic machines, there have been impediments in the form of export restrictions and certain reservations from the US and Indian sides respectively as explained earlier. As far as technical information is concerned, the General Security of Military Information Agreements (GSOMIA) is considered as one major issue. GSOMIA is negotiated by the United States with a foreign government and obligates each government to provide substantially the same degree of protection to each other's classified information. On occasion, annexes to the GSOMIA, called Industrial Security Agreements, are negotiated with the foreign government for handling classified information entrusted to industry. Program agreements (e.g., co-production) either reference the GSOMIA and Industrial Security Agreement or include security language that is substantially the same as that in those agreements. India is yet to sign the GSOMIA. But it must also be well understood that signing the GSOMIA will not automatically entitle India to all technology. It is only an understanding on how each side is to take care of the technology or information made available. From the Indian perception, any control regime automatically means restriction in the availability of or access to technology. Working their way through these issues has now become part of the process of the defense cooperation. If India and the US have a good defense cooperation going, it is no secret that avenues that were closed earlier will be opened. In the nature of the policy of both the countries it is understood that what has been established is a foundation. Both countries will have to build this up brick by brick. This of course will be possible only if the momentum is kept up by two sides and would call for a fair amount of patience and doggedness.

Attempts have been made to generally define the Indo-US defense and security relationship as a five tiered pyramid. At the base of the Indian version of the pyramid is technology transfer between two countries, topped by joint development, co-production, outright purchase of weapons and systems and at the top of the pyramid lies service-to-service relationships. The general feeling is that the American version of the pyramid is just the inverse. They have tried to emphasize service-to-service cooperation as the base of the pyramid and the top, and more difficult, as technology transfers. Though service to service cooperation is on and technology cooperation has started and is being given a shot in the arms through the JTG, technology interactions have not proceeded at the speed hoped for by the Indians and perhaps not even by US expectations. While technology interactions are desired by India and the US side too seems to be willing to meet India half way, there seems to be a lethargy in both sides doing business. However one may take heart from the words of Bruce Riedel who in a testimony to the House International Relations Sub-Committee on Asia and the Pacific on December 6, 1995, described Indo-US defense ties as being "better now than at any time in the past 30 years."

The change in Indian economy is also influencing the defense cooperation. Self-reliance and research and development continue to be the basis of the new Indian economic development policy. However, the new wave of economic liberalization has not affected India's defense industrial complex. Change has been limited to the commercial sector and has not created opportunities for American defense companies to invest in India. India needs access to the international research and development community, if its defense industry is to remain competitive. India ranks high in a number of critical technology categories according to the US Department of Defense and if she is looking for R&D collaboration, US should consider the benefits of the Indian infrastructure and resources. This would be one vital area of confidence building.

In Defense Of My Country

It appears that India's foreign policy has steadily been converging with American interests and the key to understanding US-Indian cooperation is the mindset of both parties. There is sometimes a tendency to confine Indian security perceptions unidirectionally and make comparison to neighboring countries whose concerns are small. A great deal of this confusion, misunderstanding, and suspicion derives from the fact that both countries support large bureaucracies, each of which has different agendas. One thing that India is genuinely and consistently sensitive about is the impression that Americans either wish to ignore them or attach unfair restrictions to their legitimate defense needs (Coll,1992). Americans, on the other hand, consistently complain about Indian behavior in international arenas which, if not deliberately anti-American, certainly does little to further American interests. Indian attitudes and behavior at the UN are perhaps the best example of this according to Geoffrey Kemp (Kemp, 1992). However, during the last few years, there are optimists on both sides who feel that the security relationship between India and the United States has improved and voice that defense and defense technology contacts can promote and enhance a broader political engagement.

Internationally, on the political, military and technological aspect, the US administration is extremely keen and has worked very hard to extend its cooperative international efforts. The US is eager to use defense weapons cooperation as a way of strengthening the political and military alliance with NATO countries and with Japan (DOD, 1995). Also, the US seems to have agreed to waive any possible sanctions against Israel for violating US laws in cooperating with South Africa on missile development even though Israel like India is a threshold nuclear state and not a signatory to some of the regimes. This is an interesting situation and perhaps something similar or better can be worked out between the US and India since from her point of view, India feels that she stands clear of any proliferation activity. So perhaps, serious defense cooperation with India may not be too oblique a thought and as former Indian army chief General S.F. Rodrigues said, Indo-US defense cooperation could be " a broad, blank canvas waiting for new brush strokes."

India's security concerns will always be on the agenda of her requirements and objectives in the region as she is in a difficult neighborhood with a great deal of instability all around her. Very much like the Unites States, India desires, stability in the region to continue its recent economic programs which have brought it into the global economy. There is no doubt that cooperation between the US Department of Defense and the Indian Ministry of Defense is now in the interest of both bureaucracies and should undoubtedly be pursued by both sides. Though one would like to be as optimistic as possible it would be unwise to expect too much formal cooperation on issues like proliferation because of the sharp differences in perception between the two governments.

There is enormous scope for India and the United States to cooperate in developing ideas and concepts to make the world a safer and more secure place. This requires mutual understanding and trust and a firm commitment by both nations. Any attempt to view their problems with conventional cold war blinkers is likely to produce negative results. Fresh thinking is needed on all sides with a need to engage in free, frank and constructive dialogue with reference to the realities on the ground. One hopes that is what the Indo-US defense cooperation will lead to. Looking into the future one can see an increase in interaction between the two armed forces, a greater number of visits by senior officers, joint exercises and exchange of personnel and above all an exchange of defense technology for greater security . In the nature of the policy of both the countries it is understood that what has been established till now is a foundation upon which constructive weight will be laid gradually, systematically and in keeping with the aims and aspirations of the two countries.

R. J. Augustus

13
Conclusion

The desire for defense technology, its transfer, its careful utilization, its proliferation, its management and its impact on international relations carefully looked at through the frames of reference of different nations tend to give it the characteristics of a diamond - beautiful, costly and at times deadly. The British author and physicist C.P. Snow once said, "Technology is a queer thing. It brings you great gifts with one hand and it stabs you in the back with the other." Countries aspiring to acquire leading edge technology must learn how to avoid the knife and put technology to its proper uses for as Niccolo Machiavelli has stated in 'The Prince', 'The main foundations of every state, new states as well as ancient or composite ones, are good laws and good arms....you cannot have good laws without good arms, and where there are good arms, good laws inevitably follow.' The definition of technology transfer, arms transfers and strategic exports go hand in hand and is, in the final analysis, a political and institutional question, not a technical one since products and processes are not inherently strategic or non-strategic.

When India and the US sit down at the table, there would be enough areas of conflict arising from their 'want' and 'do' lists. India's 'want' list would include the desire to be acknowledged as an emerging power, as a nuclear power and expect to be treated as one. This would have to lead to a permanent seat in the UN

Security Council. India will want the US to accept her stand on the NPT and CTBT issues as a policy of restraint rather than a challenge to the US. India will most probably rethink her policy on the NPT and the CTBT. The US must agree to accept that Kashmir is a part of India and is not under dispute. Next on India's list would be access to US high technology for her defense labs and establishments and an opening of the US market for Indian products. The US 'do' list would include their wish that India sign the NPT and CTBT, permit IAEC inspections, give up the nuclear option, stop the missile programs and weapon delivery programs, sign the GSOMIA, open up its economy on all fronts, protect US intellectual property rights and an invitation to act as a referee on the Kashmir issue. If India and the US want to have their wish lists fulfilled they would have to seriously search for common ground.

The United States and the West will define strategic exports by re-conceptualizing their security strategies during the years ahead. Clearly, a policy of export control in service of the strategy of containment will no longer suffice in meeting these security strategies. Because of the spread of high technology and the global nature of various security challenges, existing institutions for control lack the full participation of important players at the moment. The establishment of completely new regimes may be needed to take into account the concerns of all nations. Whether built on existing institutions or from the ground up, managing strategic technology in the next decade will require greater multilateral cooperation, information sharing, incentives and shared leadership.

Present control regimes built on West-East animosity are therefore not capable of formulating or implementing new security strategies. They have not been built to use technology exchange as a means of furthering security and cannot be adapted to address important security issues. A fresh look at these issues and a new arrangement has to be worked out based perhaps on some of the recommendations suggested throughout this book. The management of strategic technology will therefore require a more focused and flexible system. Technology transfer and controls must become part of a more constructive relationship with developing countries while addressing new regional and global security needs. Export control policy designed fundamentally on a strategy of technology denial to these countries must give way to a more flexible even handed policy for managing technology transfer. The policy must include key features like security trade-offs, management rather than control and a sharing rather than denial of technology. International one-upmanship, bureaucratic infighting and frantic political lobbying that has traditionally curbed effective cooperation, could prove disastrous if allowed to color technological relationships. India and the US can overcome these institutional shortcomings with strong leadership and improved government to government cooperation based on regional empathy.

India - A Time To Change

An anatomy of Indo-US relations has revealed that a new, modern, secure, and prosperous Asia can hardly be constructed without active participation by India, a large and ancient nation, that lies at its heart. As an emerging 21st century world power, India now has a robust high technological capability in most areas and is on its way to being self-reliant (Harrison, 1992). Because of this capability, the United States should not assume that India, which has maintained a good non-proliferation record, would automatically enter the technology proliferation market. However, economic pressures and Western meddling might provide a powerful incentive. While most defense analysts understand that India may be one of the countries that has need for legitimate defense technology, they also have to take cognizance of the fact that there are other countries that need the technology for less straight forward applications. So India does understand that it is better to have some restraint mechanisms than to have no rules at all. It would be even better if efforts were made to improve existing mechanisms in-order to identify real and perceived requirements and to dispel genuine concerns of recipient states.

Technology controls are spurring greater Indian self-sufficiency by encouraging expanded and accelerated indigenization. If there is no policy change, in the next decade, the Indian space and missile programs will continue to suffer multinational and national export controls. In the long run, however, these programs aided by India's growing scientific resources and technological momentum will make themselves resilient to such controls and become self-reliant. Technology embargoes and sanctions may have some near-term effect in slowing India's major defense programs, but its infrastructure has been developed to the point that it is no longer feasible for outside interests to target these programs as Indian decision makers will continue to view the program as a vital element in national security and foreign policy strategies.

To India, the MTCR is a gnawing impediment to the development of an independent launch capability and the building of a missile deterrent capability against unfriendly potent neighbors. Its major challenge is to overcome the rigors of the MTCR, other control regimes and also to withstand the political and economic pressures now being directly mounted by the major supplier-nations in conjunction with their technology controls.

An important challenge for Indian policy makers is to understand US concerns and meet them halfway without compromising the country's strategic and economic interests. Perhaps one such meeting of like minds looking for genuine solutions might occur in mid-1998. It is clear that export control regimes are driven by non-proliferation as well as an economic and technology denial motive. The US has invested a great deal in its technology base and would like to maintain its supremacy and in all fairness this is what it should attempt to do. However as long as technology export controls are used as a lever for exerting influence, the concerns of suppliers on proliferation will not be acceptable among the potential recipients. Any such policy must not only be fair, impartial and practical but also consistent in its objectives, even if they conflict with other parochial interests of the suppliers.

India is committed to non-proliferation and accepts the need for controlling exports of dual use technologies to ensure that these are used for civilian applications but reject the arbitrary, discriminatory and opaque nature of the present ad-hoc regimes, though, through its export policies, India is already respecting the guidelines of the MTCR and NSG and to some extent, the Australia Group. It is also a fact that Indian restraint, in not permitting exports of sensitive items, has been more effective than the record of many Western countries who are members of these regimes.

R. J. Augustus

An important aspect that this book has dealt with is the challenges that the controls pose for India which pivot around the ways and means by which India can gain higher returns from its defense R & D. To create a better environment to receive transferred technology, it was also emphasized that defense R & D efforts on critical technologies important to long-term national security, removing the 'red-tapism' of scientific institutions, enhancing government support for R&D, granting greater autonomy to research units, encouraging larger private-sector partnership in high-tech areas, expanding test and evaluation programs, and encouraging better inter-agency, inter-department and inter-ministry coordination and cooperation in technology development and management.

As India looks to her technological future, a quality and efficiency culture will continue to be key elements in Indian R&D. Defense production agencies should take advantage of their core competence, transforming its purely defense culture to a civil-commercial corporate structure. Looking at the defense sector alone, either on its own accord or because of peer and commercial pressure, it would be advisable for the DRDO to become leaner and hone itself into a more flexible, agile and self-reliant organization, operating in a decentralized manner to keep up with the pace of events. People at all levels will have to think outside the box and with a readiness to seize the initiative to exploit technical opportunities as they arise within the country or out. They need to understand the complexities of national security policy making so that they can help the decision-makers find ways to harness the technology revolution and better serve Indian national interests.

In a reconstruction of the Indian research organization, that this book strongly advocates, one of the key issues would be to try and leverage both the technology and production components of the commercial base as a dual-use strategy. Indian DRDO must start looking for increased access to the kind of affordable, leading-edge technology that is sustained and continuously improved through the dynamics of the commercial marketplace. DRDO labs must understand that the benefits of a better leveraged in-house industrial base are not only reduced cost, but reduced procurement time as well, which is paramount in obtaining a military advantage.

In an era of emphasis on self-reliance, there ought to be a change in the priorities associated with indigenous defense technology development. Self-reliance must be accompanied by high performance, strict program schedules and system affordability. The Ministry of Defense's policy will play a major role in the DRDO's ability to produce more affordable weapon systems. If the DRDO is to continue its planned journey, it must adopt more commercial practices. This will require a significant change in the way the DRDO conducts business with its projects and its dealing with the armed forces. Based on lessons learned, there is increasing emphasis within and outside the DRDO to use more commercial products and commercial practices in meeting military requirements rather than sticking to the old practice of doing things themselves. Although certain reforms have made a change to DRDO activities, a number of issues remain that prevent the organization from following the commercial way of business. Areas that need honing include : 1. Choosing the best contractor based not merely on the lowest quotation, but on quality and delivery schedules; 2. Long term contracts that will be an initiative to the vendor; 3. Adoption of a uniform procedure throughout all the labs and establishments; 4. Understanding of patenting laws and rights; 5. Proper identification of past problems in performance and delivery; 6. Improving the communication with other game players by removing ambiguity and increasing clarity; 7. Understanding market values of cost and pricing and finally 8. Prompt and correct payment procedures thereby building up the credibility of the organization.

If DRDO is facing problems in its projects today, one aspect of a complex solution is to bring in desperately needed efficient business practices and reduced overheads. This will not only release much needed resources but also contribute directly to the transformation of the organization's structure as envisioned by the top management. DRDO's labs and establishment will have to learn from successful corporate leaders across the country to find out how they have been able to reengineer their practices thereby putting them on par with their global competition.

One can propose a restructuring of the DRDO that should take into account :

- Closure of some of the minor labs and judiciously merging them with larger ones to avoid excess infrastructure
- Close the gap between DRDO and Defense Production
- Adopt the best business practices for all defense engineering activities, which will improve quality, cut costs and make the department more responsive.
- Subcontract engineering tasks whenever necessary while retaining design control.
- Close the gap between engineers, academicians and private defense industry by regular sabbaticals and short term postings

- Introduce contract based expertise thereby eliminating poor quality and workmanship that arises due to inertia from job security.
- Availability of electronic information and technology to all personnel in labs and establishments ensuring that they have access to state of art technical information and
- Insistence on documentation and technical and scientific publishing.

Indian efforts towards modernization and the thrust for high technology is basically founded on the premise of achieving self-sufficiency in defense technology. The crux of technology development for a developing country first and foremost calls for proper and complete utilization of the technology that has been transferred. It must be used to increase the defense effectiveness by making the most efficient use of one's collective research and development resources. There are vast amounts of resources in the military and civilian sector in India and the administration must leverage these resources. It also calls for leveraging resources through cost sharing and economics of scale afforded by coordinated research, development, production and logistic support programs. The defense forces must leverage fighting effectiveness by deployment and support of common, inter-operable equipment between the Army, Navy and the Air Force to the utmost extent. In addition, Indian efforts in the defense technology areas have to achieve a state of defense preparedness from 'just-in-case' to 'just-in-time' systems.

At the same time, DRDO labs and establishments must realize that defense technological capability alone does not generate competitive success. Scientists and engineers must acknowledge that continuous product and process improvements, organizational knowledge, commercialization strategies and other related factors also drive innovation and competitiveness in an ever changing defense industry scenario where technology life cycles have contracted, while the capital required to generate new technologies has expanded.

There have been numerous articles of criticism against the delays in DRDO projects. An analysis of why these delays occur lead one to basically three reasons. Firstly there is a question of uncertainty that is introduced by the Indian political and bureaucratic system. Secondly the funding for program should be adequate without having to scrape the barrel. Thirdly, programs like the light combat aircraft, have been undertaken for the first time in India. Dr. V.S. Arunachalam, former Scientific Adviser to the Indian Defense Minister states that surprisingly, it was not the technology delay that was holding up the light combat aircraft program for interactions with developed countries have helped in improving India's technology, but the uncertainty of a long time commitment to a program of this nature. Delays of this nature have created an impression in the West that India's national aircraft project is on a back burner. So the conclusion is that technology for the aircraft program must be restricted for it might be diverted to other programs. Such notions need to be banished about India's LCA project. India would like to assure all those pessimists and critics that this program is on the priority list of the government.

One important factor that is taken for granted in most developed countries is the question of funding for R&D. Without adequate funding, R&D is a non-starter. The R&D funding scene in India is enough to demoralize the most ardent researcher. India's integrated guided missile program, the light combat aircraft and the main battle tank projects have yielded useful high-technology results notwithstanding the delays. However at times there has been a danger that the momentum of the indigenous technology development might be sacrificed. While there has been some inability of India's industry to absorb the high-technology developed by DRDO's scientists and engineers, Indian leaders and lawmakers have not been able to fully understand the teething problems of indigenous development of technology. So programs which required sustained government commitment received it only in fits and starts. However, lessons have been learned and things seem to be improving over the last couple of years.

The DRDO is also partly to blame for the three factors leading to the delay. The scientists of DRDO do not have an appreciation of realistic time frames for their projects. They do not have an appreciation of the financial support required for such projects. They do not have an appreciation of India's bureaucratic juggernaut. Not having undertaken such major projects before, they are only now getting a handle on the issues related to indigenous technology. Those who decry DRDO's indigenous development have chosen to lose sight of these major factors. User satisfaction in the Indian environment is a major problem. Users of the DRDO equipment are sure that if they are not satisfied with the performance of the indigenous equipment, they could always purchase equipment from outside the country. There have been cases of imported equipment having far more problems than the indigenously developed system. If the users in developed countries had followed the same thought process the research and development establishments and defense industry in those countries would not reached the level they have today. These establishments have enjoyed the constant undiluted support of their users. Of course, there has been a gradual change in attitudes in India over the last few years. Developed countries have gone through the industrial revolution and know what it takes to develop defense technologies. Their users also take tremendous pride in their national products and part take in its development, without the attraction of going out of the country.

From the Indian perception, any control regime, besides the moral principle of discrimination and other things, automatically means restriction in the availability or access to technology. India should therefore join some of the international regimes, not necessarily with an eye on immediate gains, but for the benefits over a longer time span. India may consider it as another Caesar's coin. That, of course, is part of the process of cooperation. By pursuing overall cooperation with the US in defense technology, from the political point of view, it would help strengthen the bonds between India and the west. From the military perspective, an increased likelihood of operating in a coalition environment means the West will stress inter-operable equipment and rationalized logistics and India will be able to quickly synchronize and participate. Economy forms the third reason for defense budgets of all countries coming down and no single nation can individually afford many of the large developmental programs. Fourthly, a defense cooperation will enable India to utilize western test facilities and this will endorse her indigenously developed technology, thereby making it more market worthy. Sharing of research costs, access to innovative foreign technologies, achieve economies of scale in production, entry into international markets and technology awareness are some of the added benefits.

In line with this doctrine there would then be the need for Indian R&D to right-size its infrastructure; reduce the cost of defense systems ; implement acquisition reform and try and leverage the industrial base of the technologically developed countries. When Indians look at the technology available, its transfer, opportunities for cooperation, opportunities for modernization and the overall development of the national technology base, they will have to weigh all the pros and cons and choose a path that will put India at the leading edge of technology. In the words of Robert F. Kennedy, "Some look at things as they are and ask, Why? Others look at things as they could be and ask, Why not?" India should undoubtedly ask "why not?"

US - A Time To Rethink Policy

The year 1990 noted a significant change in export control strategy even if it was not a complete overhaul. The US was forced to acknowledge that global economic competitiveness is vital to her national security. It was also forced to realize that export expansion could be an enhancement and not really a threat to security. On the other hand, a system of extensive controls would stifle the civilian technological advancement essential to maintain a strong military, economy and international harmony (Long, 1992). The changed political scenario opens up new horizons for technology transfers that could promote economic reforms in developing countries and perhaps even lead to democracy in some. This could also prove to be a more effective way of enhancing security strategy rather than a policy of denying technology.

American policy towards technology transfer will still recognize potential risks but due weightage will be given to the two factors of 'who is asking?' and 'what is being asked?' While there are bound to be trade-offs, the US must realize that they can gain much through cooperation and also that the strategic value of technologies lie not only in controlling their transfer, but in sharing them. So the US administration will find that technology denial does carry a price and will therefore need to evaluate the costs and benefits of transfers. One priority of the US will be to work towards Indian participation in multilateral technology regimes sending a clear message that membership is not restricted to the West.

With its undisputed superiority in high technology, America was confident about its business opportunities and support of its allies. Today each country strives for its own economic survival and do not always look to the US for guidance. The Asian suppliers are moving into areas that were predominantly American turf and the US government cannot deny her high-technology industry the right to compete if she wants to maintain the technological lead. Americans are learning that technology transfer is no more a one-way street. Again, it is becoming extremely difficult to draw a line of demarcation between military and civilian technologies. Any attempt by the US to restrict access to high technology is resented as a kind of discrimination. Since most developing countries consider conventional weapons as a legitimate means of self-defense, the US must rethink cooperation in technology development. These countries are apt to remind the US that science and technology, research and development are birth rights and represent freedom. What the research leads to is another question.

In recent times, India and the US have made efforts to resume a strategic dialog. The Defense Policy Group and the Joint Technical Group address issues that were skirted in past years. Reciprocity in all interactions is the order of the day and any bargains driven must be advantageous to both countries. There has been a marked change in US policy, at least from what Indo-US meetings and talks suggest. But will it filter down to the working level ? Will it ease export restrictions ? Export control regimes cannot survive in a vacuum. They are possible and effective only under circumstances where threats and opportunities are clearly defined. However export controls on dual-use goods, technology, and software will continue to be one useful tool in US efforts to relieve strategic concerns by stemming proliferation and to block or slow down the emergence of new defense technology suppliers, provided the US learns how, when and on to whom to apply it. A blanket policy on dual use items will be a non-starter. Technology transfer requests will have to be endorsed based of course on complete transparency of its intended use. This would be most helpful to countries like India that have need for components and equipment for exploratory development, but are unable to now import it from the US. The implementation of such an idea may not be that simple. The US would like to use export controls to maintain the superiority in leading edge technology and to delay the transfer of critical defense technology. But relaxing security controls will promote commercial and technological assistance to developing countries and at the same time fill the coffers of the US industrial houses. The dilemma is seen in the recently announced US Arms Transfer Policy which suggests endorsing certain transfers in-order to "enhance the ability of the US defense industrial base" and suggests restraints on transfer to "preserve its military edge." Such conflicting goals are bound to compound the task of formulating any international regulations.

    The premise of US policy has been that most countries do not have legitimate reasons for possessing nuclear weapons, or at least their reasons are not as legitimate as those that motivated the United States to develop nuclear weapons. With most of the Chinese long range missiles aimed at US cities (Gertz,1998-B), perhaps the US does have something to think about. But what then would be India's plight? The US will have to consider the consistent decades old Indian security concerns in view of this revealed threat. The cornerstone of US nonproliferation policy during the last several decades was to prevent the spread of high technology that might help countries to achieve this capability. India as an emerging world power, deserves that her strategic policies be considered more seriously by the industrialized countries. Because of her technical capability, the US might feel that India could also export sophisticated technologies, thereby creating proliferation concerns. This would then qualify India as a prime candidate against whom export controls must be applied. But a number of countries have the expertise required to develop weapons on their own if they choose to do so and technically, neither the United States nor any other country is in a position to condemn another country's decision to possess high technology. US policy does exactly this and developing countries do not think that the United States is the best judge of their requirements. It is difficult to sell this idea where conflicting ideas of sovereignty remain. One must realize that a platform made up of the 'Haves' and the "Have-nots" is not an ideal one on which to attempt building a secure and sane world order.

If there is no other option other than nuclear states coexisting with one another then the US needs to accept this fact. The principle that all states have the inherent right to possess dual use high technology must also be accepted. Given the trend of the new American diplomacy, one can be fairly certain that a way will be found to make the Indo-US dialog a win-win situation. If unfortunately the leaders on both sides think of isolating themselves, then golden opportunities would be lost for America and India.

Finding Common Ground

What is the give and take ? The United States wants to maintain its dominant place in the international system and it wants to influence international order. India wants to get out of its present rut and feels the need to be recognized for its adherence to democracy and as one of the world's oldest civilizations. Selig Harrison (Harrison, 1997) says that 'India is attempting to assert major power status without incurring the economic and diplomatic costs that overt weaponization would involve'. Harrison talks of a nuclear 'bargain' wherein India would retain its nuclear weapons options but would have to agree to a series of concessions. Would Indian policy makers accept Harrison's three concessions ? Firstly, signing the nuclear treaty or stop testing without signing the treaty; secondly submitting to international safeguards and thirdly India would have to make a binding commitment not to export nuclear technology which is what India follows anyway. It is understood that 'influential analysts and highly placed officials are quietly discussing it'. A positive outcome would be beneficial to Indian defense industry and a shot in the arm to those idealists who view the transfer of technology for technology's sake.

Senator Sam Nunn says that the United States is not playing a very big role in South Asia (Nunn,1997). But the US administration has been preoccupied with the Indian nuclear program for years and has applied export restriction on exports to

India. India in turn has restricted her relations with the US, creating a lose-lose situation. In today's political setting this is absurd or at the least unproductive since the interest of both countries lie in enhancing political, economic, military cooperation. Pressurizing India on the nuclear issue is one sure way to alienate the country. As explained earlier, India is a mature country that knows how to handle its technology, nuclear or otherwise as history has shown. The US would do well to help India in her search for civilian nuclear power technology as is being envisaged in the case of nuclear and missile potent China and against some US concerns. Secondly the US must think of India as India and Pakistan as Pakistan and try not to link the two countries in their strategic appraisal. The US should form strategic partnerships with each country with no Pakistan or Indian strings attached when dealing with the other. Thirdly, the US must go by the 'Agreed Minute' on Indo-US defense cooperation. As in most cases, what is decided at the topmost level fails to filter down to the working level. If the US feels that some form of restriction on its exports must be applied, then it should find a more rational way of going about it. The present arbitrary application of export restriction seems to underline the fact that what has been decided in the top echelons of the US administration are not known to or interpreted correctly by those working the daily issue.

The United States insists that it needs nuclear deterrence for security and would like to retain nuclear monopoly while India's own security is affected by her nuclear potent neighbors. The United States, says Kemp (Kemp, 1992), must accept that India herself is the best judge to think through the pros and cons of proceeding with her nuclear and missile programs at a time when economic considerations are critical in today's competitive world.

In Defense Of My Country

So the United States can help stimulate a serious debate about the merits of arms control within the Indian decision-making process without undermining India's security prerogatives.

The idea of any Indian missile targeting the United States is too farfetched. So it is difficult for the US to argue that the she feels threatened by the Agni or any other Indian missile program. But one must concede that a technologically successful India could become a potential missile exporter, depending on the circumstances. Since there is certainly some truth in this, US interests will be best served by coaxing India to restrict Indian missile technology exports, if any, rather than taking the futile step of trying to stop the Indian missile program outright. Unfortunately, much of the American rhetoric aimed against the Indian missile program has been in trying to force India to follow control steps it does not consider to be in its best security interests (Ali, 1991; McDonald, 1991).

Americans feel that one confidence building measure to get Indian cooperation would be to work out a deal for US high technology in exchange for Indian formal compliance with MTCR guidelines notwithstanding her following the guidelines on her own accord. Getting India to agree on more rigid controls of its own exports and cooperation with other countries would be in the US interest, particularly since it would provide strong leverage for the US to use the same guidelines and approach for dealing with Pakistan and China.

Countries in the subcontinent, with sufficient financial, industrial and human resources and determination can in the long run, in a variety of ways, procure weapons systems they desire. So the most effective means of keeping certain categories of weapons out of a region is to convince all the countries in the region to abstain from acquiring such systems. The US must be prepared to promote, help negotiate, monitor, and participate in regional arms control undertakings compatible with its national security interests but should not seek to force these obligations against the wishes of affected countries. After all, countries in this region will definitely listen to clear logic put across to them without any hidden agenda. Carter (Carter, 1992) suggests that a cooperative form of security offers the best prospect of addressing new international problems and the development of a regime for cooperative engagement is the new strategic imperative. He further clarifies that cooperative security is designed to ensure that organized aggression cannot start on any large scale and to combat proliferation, it is absolutely imperative to secure the concurrence of all the advanced nations in any proliferation control strategy that can possibly be effective.

For India and the US, economic growth will be increasingly driven by technology-intensive businesses. From where India comes, technology is undoubtedly the most important ingredient in the formula for stable and sustained economic growth, higher living standards, and global competitiveness. From the US perspective, the high technology industries that are driving its economic growth are expanding and changing dramatically. With international tie-ups, technology development, production and sourcing have become more complex and challenging. With the mushrooming of 'multinationals', research and development has become much more global. These 'multinationals' that have moved into India and are pursuing research and development strategies and developing new technologies. Apart from raising levels of technology, employment and living in India, they are also contributing to US technological capabilities One must realize that the U.S. which was at one time responsible for about 70 percent of the world's R&D has now come down to 30%. By joint research, development and co-production, US firms can adapt products and production processes to local markets while tapping local sources of expertise. With the emphasis to reduce R&D costs, US firms are reorganizing their R&D operations, and applying new performance standards and contracting out increasingly large portions of their R&D to sources overseas. This could very well apply to defense technology interactions between the two countries as well, taking into consideration all its associated precautions for the national security of both countries.

There should be a transparency in the relation so that both sides realize that the other is trying to play fair. There should be no 'buts' and 'or elses' in the interaction, but both countries should work around the areas of conflict, without getting into the same rut that they have been in for a very long time. There are optimists on both sides who are paid to be pessimists in the interest of their country's national security. While accepting this, more creative thinkers are needed at the working level ; those who can go beyond the cosmetics of voicing their respective administration's age old and time worn policies. After al,l a ship is safest in the harbor, but that was not what it was meant to do. India will have to accept the US as the world's most powerful country and the US will have to make India a foreign policy priority.

The United States is seen as accommodative to the Israeli and Pakistani nuclear ambitions while being unmindful of India's legitimate security concerns vis-à-vis China. Surely those more knowledgeable are aware that it would hardly suit the US national interests to have a conventional India facing a nuclear China in collaboration with a nuclear Pakistan. Gradually, while Indian policy makers are becoming aware that the United States is in the process of reassessing its relations and policies, there is a hesitancy at every step. Bureaucratic and procedural red-tape is reeled out to mask the 'play-safe' attitude. What is lacking is the conviction and the courage to throw of the old mantle of suspicion and forge ahead with new decisions backed by all of one's integrity.

There have been major changes on the American and the Indian side on technology, technology export and import related issues. The impact of the new policies on Indo-US commercial relations is likely to be felt in the coming years. In the area of defense technology transfer, what has been attempted is to bring in a transparency to the pertinent issues and opportunities, the ways and the means for facilitating the transfer of technology and catalysts for setting up working equations between India and the US. The analysis carried out for technology transfer drivers, the transfer mechanisms, the barriers to technology transfer and models of Indo-US interactions that have been developed specifically for the Indo-US context, are an attempt to bring in a clarity to the technology transfer problem that other third world countries might also face when attempting to deal with their western counterparts. The development of a model for the Indian methodology of absorption and management of technology in-house represents the Indian R&D system that is being revamped.

While some aspects of this analysis have negative aspects concerning the ready resolution of US-India bilateral issues relating to technology transfer, there are also some positive aspects to the problem as well. Extracting a positive outcome requires some creativity and flexibility on the part of Indian authorities, as well as a willingness to include specific technology transfer issues on the senior bilateral political agenda as well. More than half a century of experience with the munitions licensing system ought to have prepared the US to deal with the export control situation they have with India. So one can always hope that they will come up with a solution to mutual advantage.

Although there has been a certain unevenness in the Indo-US relationship, there is a strong sense of agreement on those broadly-defined goals between the two countries and a partnership does require a certain amount of pushing and pulling. It is a partnership that just might provide the direction into the future provided the two countries do not allow any single issue to hold the relationship hostage, however important it is, but should attempt to put things on an even keel. US strategy in the region is being reworked both because of domestic factors and new challenges that are emerging. In the new equations, the United States, India Japan and China, must be forces of peace and stability in the region. Policies that promote cooperation, balanced force levels, non-use of force, arms control, and confidence building, are the needs of the time. More philosophically this author feels that the difference between a CBM (Confidence Building Measure) and an ICBM (Inter Continental Ballistic Missile) is the 'I' which is nothing but the ego of nations and individuals involved. Once the ego is removed or curbed, defense and defense technology international relations would take on a very different connotation. It is hoped, in the words of Jefferson' that 'our principal men be men of principle'.

The author of this book is not so naive as to believe that adoption of all the recommendations will lead to a Indo-US equation where nuclear capabilities will be accepted and Western high technology needed for the design of advanced weapons will be freely available to India. This has never been the premise of the concept of defense technology transfer as far as India or this author are concerned. The transfer of legitimate technology for improving the standard of R&D organizations, the standard of technology and the standard of living has been the corner stone of all arguments.

Of course there will be certain exports to the region, but a vital part of the equation for Indo-US cooperation depends upon at least increasing or hastening the flow of authorized US technology. As we move into a new phase of export controls in light of the changing political and economic landscape, a streamlined, more narrowly tailored enforcement structure offers increased merit and attention. There is the range of opportunities being presented as a result of more mature thinking on both sides. Apart from similar ideologies, the economic relationship offers a particularly strong means of bolstering overall common goals and values that characterize the relationship at its best. Americans need to regain some of their pioneering zest for taking risks and Indians need find ways to expand incentives for Americans to take those risks.

Though a number of inputs to the Indo-US working equation have been identified one does not know how this will stabilize, whether it is going to lead to a formal alliance or just better understanding or a lot more cooperation as the human factor in relations will always come into play. There has certainly been a sort of sea change and the relationship has moved from the valley to the hill and the view from the hill is definitely better. However there are bigger hills to climb in the near future and the view is bound to get better. India and the United States do have a unique door of opportunity wide open to them to establish a stable and secure relationship. The future is out there waiting and the United States and India have the opportunity to define that future.

In Defense Of My Country

R. J. Augustus

In Defense Of My Country

Addendum

On 11 May 1998, the feast of Buddha Purnima, India carried out three nuclear tests. Two more followed in close succession.

This book was initially submitted and accepted as a Ph.D. dissertation in the fall of 1996. It has been minimally revised up until early May 1998. Before it could go to the publishers, the five Indian nuclear tests took place. A few of the statements in this book have been blasted away with the tests but the major part remains intact and much more meaningful in the present situation in which India and the US find themselves.

India has gone the French and the Chinese way and Pakistan threatens to follow suit if India is not 'punished' or is 'left off the hook'. The debate rages whether it was true security concerns, domestic pressure brought on by the testing of the Pakistan Ghauri Missile or an agenda in the Indian election manifesto. Whatever be the reason, the deed is done and the US administration has ordered sanctions on India. There is a blind international clamor to make an example of India. There are those who are heaping abuse on India and also trying to push the United States into a corner. But sanity and wisdom still prevails if one goes by the editorials, articles and talk shows appearing in the media.

Among the cooler heads, it is acknowledged that India has done what she had to do and the United States has done what she had to do. Now, however is the time for the US and India to sit down and rewrite all the equations that were talked about in this book. Sanctions on the part of the US and stubbornness on the part of India will only escalate the tension and throw away years of hard work by both sides towards improving overall Indo-US relations and personally for the author in the area of Defense Technology Cooperation. It is hoped that the trust carefully built up over the last few years is not blown away like sand in the desert.

Again it is hoped that 'our principal men be men of principle'

Rayol John. Augustus

18 May 1998

In Defense Of My Country

Appendices

APPENDIX 1

Form To Be Submitted By The Applicant To The Import Certificate Issuing Authority For Issue Of Import Certificate (IC)

Under The Indo-US MOU

1   (a)   Name of the Undertaking with address of Head Office
      (b)   Location of the factory
      (c)   Name and address of US Exporter
2.   Name of the sponsoring Directorate in ICIA
3.   Items of import for which import certificate is required
      Details of the item   Quantity   CIF Value

4   If required for manufacture
(a)   Details of the industrial license/SIA Registration / DGTD Registration.
(b)   End Product as given in the Industrial License
(c)   Actual item of manufacture
5.   If required for R&D
(a)   Details of the Registration with D S T along with validity
(b)   Specific project for which the item is proposed for import
6   (a) Is the item given in 3 above
      C G or a part thereof?   Yes/No

If Yes

(b) If CG is under non-Open General License (OGL), details of Import License number & date along with photocopy of the License with list indicating the serial number of the item sought

(c) If CG is under OGL the Import Policy (IP) Reference in the Tables

(d) Name of, the Licensing Authority with full address which has issued the Import License

7. (a) Is the item given in 3 above- a component/raw material ? Yes/No

If Yes

(b) If under OGL, the classification in I P

(c) Details of the IP under which import was allowed (enclose copy of the letter of approval along with list)

(d) If non-OGL. details of the Import License number & date under which permitted, along with a photocopy of the License and Debit Statement of Customs

(e) Name & full address of the Licensing Authority which has issued the Import License

- Copy of the letter from the US exporter in support of the request for IC and the reference number of the Controlled Commodities/Munitions List of the US Export Administration Regulations
- The items being imported will / will not be integrated into Indian end-products to be re-exported.

10. in-order to import the item(s), I

(Name) ....
.authorized by

Name of the organization.... . .... a Govt. entity/Govt.-controlled entity / private ....

sector entity hereby undertake to import the item into India and not to re-direct it, or any part of it, to another destination before its arrival in India; to provide, if asked, verification that possession of the item was taken; not to re-export the item without any written approval of the Import Certificate Issuing Authority; not to re-transfer within India the item(s) specified in this Certificate without the written approval of the Import Certificate Issuing Authority; to obtain permission in writing from the Import Certificate Issuing Authority prior to any change in end-user which shall be preceded by the new end-user notifying the import Certificate Issuing Authority that it agrees to the conditions contained in this document.

Signature

Date   Designation

To be signed by the authorized representative.

APPENDIX 2

Form To Be Submitted By The Applicant To The Import Certificate Issuing Authority For Import Certificate Issue

Under Indo-US MOU For Higher Assurances

1. Description of the item (indicate IC and whether item is raw material / component or end-product)
2. Quantity
3. Value (CIF)
4. Name & address of
a) Manufacturer
b) Exporter
c) Supplier
5. The items being imported will/will not be integrated into Indian end-products to be re-exported.
6. In-order to import the import items, I Name …….………………..….. duly authorized by Name of the Organization …………………………………………… a Government entity/Government-controlled entity / private sector entity / entity covered under ………………………………………………….. -hereby undertake to-provide special security for the items listed above while en-route to India and special security for the item(s) while in transit and within India. Special security arrangements made would be approved by the concerned Import Certificate Issuing Authority (ICIA) prior to the item(s) being released from Customs.

In Defense Of My Country

-to ensure that access to the items listed under 'Description of Goods' in the Import Certificate is permitted only to duly authorized Indian nationals and representatives of the country of origin. For nationals of third countries and / or their representatives of whatever nationality, I shall obtain permission from the Import Certificate Authority before access to item is granted.

Date   Signature of authorized representative

                                              SF-AL/STAMP

Place

Name

APPENDIX 3

Steps in the Import Certificate (IC) Procedures for Government / Government-Controlled Entities
(Read along with Figure 27)

1. Indian importer and US exporter reach agreement on the sale (Purchase Order).

2. The US exporter tells the Indian importer that an IC is required in-order to get an export license from the US Government. The US exporter would detail the item(s') for which the IC is sought and indicate the reference number in the Controlled Commodities / Munitions List of the US Export Administration Act.

3. The Indian importer applies to the appropriate ICIA giving all details of items to be imported including their description & approximate values(Format in Table 12).

4. ICIA issues an IC, to the Indian importer and sends a copy of it to the Ministry of External Affairs.

5. The Indian importer sends the original IC to the US exporter.

6. The US exporter applies for an export license to the appropriate US Government Agency.

7. The export license is issued, if extra assurances are not needed.

8. If the US Government feels that extra assurances are required for selected items, the US Government requests the Ministry of External Affairs for extra assurances.

In Defense Of My Country

9. The Ministry of External Affairs forwards the US request to the ICIA for assessment.

10 The ICIA processes the case and conveys its decision to the Ministry of External Affairs.

11 The Ministry of External Affairs conveys or denies the extra assurances to the US Government, and

12 Intimates this decision to the Indian importer.

13 The appropriate US agency then issues or denies the export license to the US exporter.

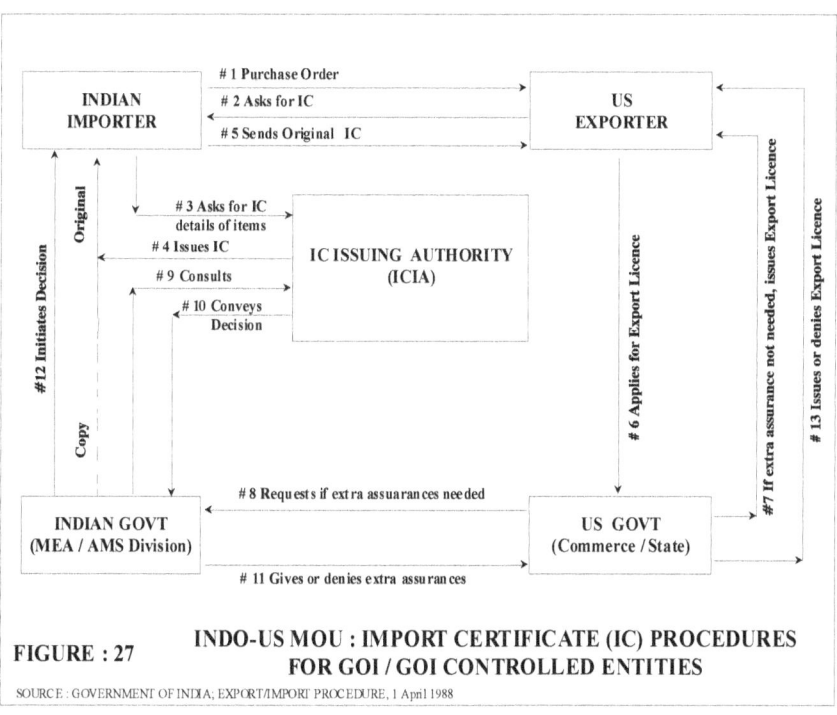

FIGURE : 27 INDO-US MOU : IMPORT CERTIFICATE (IC) PROCEDURES FOR GOI / GOI CONTROLLED ENTITIES

SOURCE : GOVERNMENT OF INDIA; EXPORT/IMPORT PROCEDURE, 1 April 1988

APPENDIX 4
Steps in the Import Certificate (IC) Procedures for Private Sector Entities
(Read along with Figure 28)

1. Indian importer and US exporter reach agreement on the sale (Purchase Order).

2. The US exporter tells the Indian importer that an IC is required in-order to get an export license from the US Government. The US exporter would detail the item(s) for which the IC is sought and indicate the reference number in the Controlled Commodities / Munitions List of the US Export Administration Act.

3. In case of items not under Open General License (OGL), the Indian importer will obtain, a license for the import separately from the concerned licensing authority of the GOI before applying for an IC. In case of items under OGL, the Indian importer applies to the appropriate ICIA giving all details of the items to be imported, including their description and approximate value (Format in Table 12).

4 ICIA issues the IC and forwards it to CCIE.

5. CCIE issues an import License if the item is under OGL and forwards it along with the import Certificate to Indian Importer. A copy is sent to the Ministry of External Affairs. In case of non-OGL items, the importer will present his Import License to CCIE who will endorse the MOU assurances on the Import License for controlled items, to be imported and forward the same along with the IC to the Indian importer and copy to the Ministry of External Affairs.

6. The Indian importer sends the original IC to the US exporter.

7 The US exporter applies for an export license to the appropriate US Govt. agency.

8. The export license is issued, if extra assurances are not needed.

9. If the US Government feels that extra assurances are required for selected items, the US Government requests the Ministry of External Affairs for extra assurances.

10 The Min. of Ext. Affairs forwards US request to appropriate ICIA for assessment.

11 The ICIA processes the case and conveys its decision to CCIE .

12 If it is decided to give extra assurances, CCIE recalls the Import License, issues an amended Import License and sends a copy to the Ministry of External Affairs.

13 The Min. of Ext. Affairs conveys or denies the extra assurances to the US Govt.

14 The appropriate US agency then issues or denies export license to US exporter.

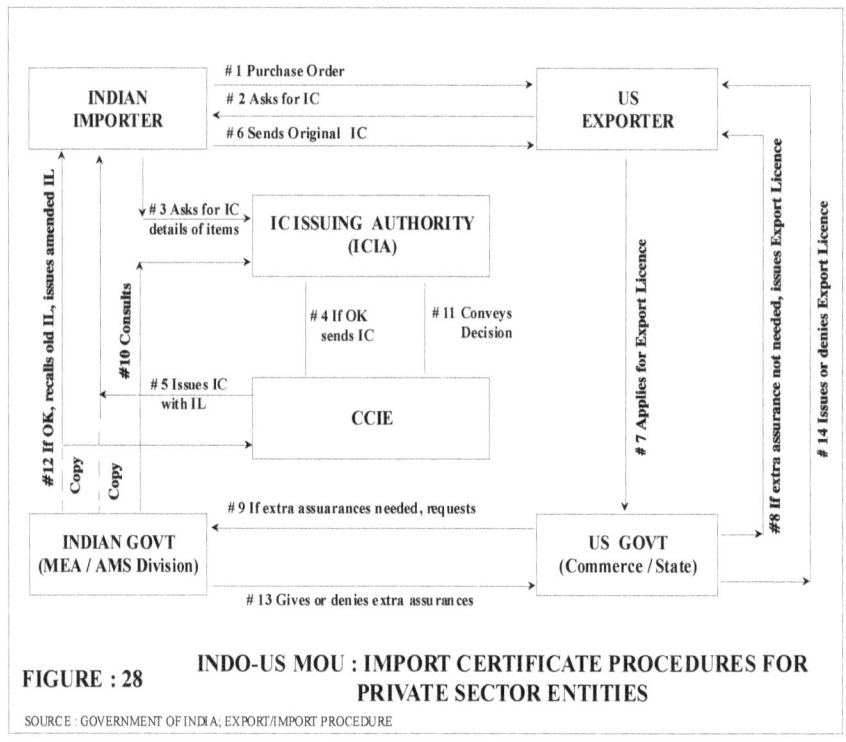

FIGURE : 28   **INDO-US MOU : IMPORT CERTIFICATE PROCEDURES FOR PRIVATE SECTOR ENTITIES**

SOURCE : GOVERNMENT OF INDIA; EXPORT/IMPORT PROCEDURE

APPENDIX 5

Delivery Verification Procedures

1. Simultaneous with issue of the US Government export license, USG will inform US exporter that Delivery Verification (DV) is required. The exact details of the items for which DV is required will be given.

2. US exporter informs Indian importer accordingly, well before the shipment is to arrive in India.

3. Indian importer procures blank DV Certificate set of forms from concerned ICIA and presents the same, duly filled in, to the Customs authority in triplicate. The Customs authority at the point of entry will compare the DV Certificate with importer's Bill of Entry and certify that the import has been effected into India and that the commodity is either released for home consumption or ware-housed.

4. Indian importer forwards one Copy of the DV Certificate to US exporter and another to GOI (Ministry of External Affairs). The third copy will be retained by the Customs authority along with the Bill of Entry relating to that particular import.

5. US exporter presents DV Certificate to USG licensing office.

R. J. Augustus

References

Aaron H.J., Collins S.M., Bryant R.C. and Lawrence R.Z. Preface to the Studies on Integrating National Economies ; Techno-Nationalism and Techno-Globalism - Conflict and Cooperation by Ostry S. and Nelson R.R., The Brookings Institution, 1995.

Abraham I. "India's 'Strategic Enclave' : Civilian scientists and Military Technologies." in Armed Forces & Society, 1 January 1992, Vol 18, p. 231.

Abrams E. "Security and Sacrifice : Isolation, Intervention and American Foreign Policy"; Hudson Institute, 1995

ACDA - US Arms Control and Disarmament Agency, Office of Public Affairs- Press Release ; Washington D.C. 20451, October 20, 1995

Adler F. "International Technology Transfer"; ARL's International Technology Transfer Program, International Technology Transfer Branch, U.S. Army Research Laboratory; AMSRL-TT-TI. Posted on the Internet, July 1996.

AER - "Missile Technology Control Regime", Arms Exports Regulations; Oxford University Press, 1991, p. 226.

Ahearn R.J. "Super 301 Action Against Japan, Brazil and India : Rationale, Reaction, and Future Implications"; CRS Report for Congress, 26 January, 1990

AIA - Aerospace Industries Association Newsletter No. 5 December 1994

Albright D. & Hibbs M : ' Iraq's shop-till-you-drop nuclear program'; Bulletin of Atomic Scientists, April 1992

Ali, S. and McDonald H. "South Asian NFZ?" Far Eastern Economic Review, December 5, 1991, pp. 20-22.

AMR Asian Military Review - Regional News Update; Volume 4 - Issue 1, February / March 1996 p. 82.

Apotekar A., Mills S. and Palese B. "The Comprehensive Test Ban Treaty in Jeopardy?"
Armed Services Committee on February 7, 1995.

Arnett,E. "The Comprehensive Test Ban Debate." Issue Paper; AAAS Program on Science, Arms Control, and National Security; AAAS Publication No. 88-7 1988

Arora C.K. 'Clinton Urged to back India's Claim to UN seat'; The Sunday Pioneer, 15 March 1998.

Arquilla J. "Between A Rock and A Hard Drive: Export Controls on Supercomputers." The Non-Proliferation Review ; Winter 1996.

Bajpai U.S. "India and its Neighborhood"; Lancer International, Delhi, 1986.

Baldwin D.A. "Power Analysis and World Politics"; Power Strategy and Security; Ed. Klaus Knorr, Princeton University Press, 1983. pp. 3-36.

Banerjee D 'India and its Strategic Options'; The Pioneer, 26 March 1998

Barnard R.C. " India Offers Huge Business Potential." In <u>Defense News</u>, March 20-26 1995. pp. 31-32.

Barnes Harry G., Jr. "The United States and India: The Economic and Technology Dimension" at a conference on <u>India and America After the Cold War</u> co-sponsored by the Carnegie Endowment for International Peace and the India International Center, New Delhi, March 7-9, 1993.

Bernstein D. "Conversion; Can the Russian Military-Industrial Complex be Privatized ?." Ed. Michael McFaul, <u>Stanford University</u>, Center for International Security and Arms Control; 1993.

Bhagwati J. "India and the United States - Issues and Opportunities." Talk at the <u>Embassy of India</u>, Washington D.C, March 22, 1994.

Bidwai Praful " New Delhi is Wrong to Reject Proposed CTBT." <u>India Abroad</u>, 16 August 1996, p. 2.

Blackwell R. D. and Carnesale A. " <u>New Nuclear Nations - Consequences for US Policy"</u> ; Council on Foreign Relations Press, New York, 1993.

Boyer Y. "Conventional and non-conventional weapons : Their changing characteristics and challenges for arms control." In <u>The Defense Trade : Demand, Supply and Control</u>, Volume III, Royal Institute of International Affairs, 1994.

Brahma Chellaney "The Global Diffusion of Military Technology." <u>In The Proceedings of a Workshop held at the University of Wisconsin</u>, Madison, December 6-8, 1991 (Madison, WI: Center for International Cooperation and Security Studies, University of Wisconsin), p. 19.

BXA Bureau of Export Administration, U.S. Department of Commerce, posted on the Internet 10 October 1995.

Cardomone T. A. "The Worldwide Arms Trade: Potential For U.S. Leadership"; Conventional Arms Transfer Project; A Project of Council for a Livable World Education The prepared text of a speech delivered to the James A. Baker III, Institute For Public Policy, Rice University - Houston, Texas, October 23, 1995.

Carter A.B., Perry W.J. and Steinbruner J.D. "A New Concept of Cooperative Security"; The Brookings Institution, Washington D.C. 1992. p. 2.

Carr R.K. "Measurement and Evaluation of Federal technology Transfer"; Proceedings of the 20th Annual Meeting of the Technology Transfer Society, Washington, DC July, 1995.

Caruana P.P. "New Opportunities for Partnership: The Military Vision." Remarks to the
Armed Forces Communications and Electronics Association, Rocky Mountain Chapter, Colorado Springs, Colo., 13th Annual Space Symposium, August 2, 1995.

CCSTG; "New Thinking and American Defense Technology"; Carnegie Commission on Science, Technology, and Government ; 1993.

Chalmers M. and Greene O. and Wulf H. "The United Nations Register of Conventional Arms." In SIPRI Yearbook 1993: World Armaments and Disarmaments (Oxford University Press, 1993), pp. 533-44.

Clapper J.R. Jr. "'The Worldwide Threat to U.S. Interests." Prepared statement to the Senate Armed Services Committee; Defense Issues: Volume 10, Number 5, Jan. 17, 1995.

Cole J. and Cooper H. "Buyers of US Arms Toughen Demands." In The Wall Street Journal, 16 April 1996.

Coll S. "India Faces Nuclear Watershed," Washington Post, March 7, 1992, pp. A15 and A19.

Cooper K. J. "Indian and Pakistan Cold War Shifts to Nuclear Matchup -Arms Race Raises Stakes in Long-Standing Rivalry", Islamabad, Pakistan, 1996.

Cyranek G. "Information Technology for Development" in Technology Transfer for Development -The Prospects and Limits of Information Technology. (Summary by Ansen D. and Kells K.), in Rueschlikon, Zurich, Switzerland; 5 December, 1991.

Czerwinski T.J "The Third Wave: What the Tofflers Never Told You." In Strategic Forum ; National Defense University, No: 72, April 1996.

Damodaran A.K. "Extra-Regional Interests; . India and Its Neighborhood"; Ed. US Bajpai; Lancer International, Delhi, 1986; pp. 372-386.

Dean J.G. Ambassador's preface in "A Common Faith- 40 years of Indo-US Cooperation, 1947-1987" ; USIS, 1987.

Delgrego W.J. "The Diffusion of Military Technologies to Foreign Nations"; Thesis presented to the School of Advanced Airpower Studies, Air University, Maxwell Air Force Base, Alabama, June 1995

DN-1 "Sanction China Now"; Commentary, Defense News, 5-11 August 1996, p. 12.

Denman G. L "Defense Initiatives in Technology-Based Partnerships." Remarks by Director, Advanced Research Projects Agency, to the Technology-Based Partnership Conference, Santa

Barbara, Calif., : Defense Issues: Volume 10, Number 16; Feb. 2, 1995.

Deshpande G.P. "China in the South Asian Context." In India and Its Neighborhood; Ed. US Bajpai; Lancer International Delhi, 1986. pp. 359-371.

DFAW "Pakistan Takes Delivery of First Home-Built Tank and APC," Defense and Foreign Affairs Weekly, April 2-8, 1990, pp. 2-3.

DISAM A Comparison of Direct Commercial Sales and Foreign Military Sales for the Acquisition of US Defense Articles and Services; DOD Defense Institute Security Assistance Management : GPO, 1993.

Dixit,A. 'A New Foreign Policy on the Anvil'; The Pioneer; 18 March 1998

DOD - "Defense Science and Technology Strategy" ; Department of Defense Director, Defense Research and Engineering, September 1994 .

DOD - "Dual Use Technology - A Defense Strategy for Affordable, Leading Edge Technology"; Department of Defense, February 1995.

DOD - "Proliferation Threat and Response"; Department of Defense, Pentagon, 1996. pp. 37-39.

Donowaki M. " Developing a code of conduct for conventional weapons" Non-Proliferation Review, Fall 1995.

Dorn E. "10 Ways to Look at the Department of Defense"; Remarks to the 26th annual Student Symposium sponsored by the Center for the Study of the Presidency, Washington,

: Defense Issues: Volume 10, Number 36. March 25, 1995.

Drace-Francis C. "From Policies to specific decisions: Practice and Problems." The Defense Trade: Demand supply and control. Volume 3; Royal Institute of International Affairs, London, 1994.

Dunn L.A. "It Ain't Broke Don't Fix It"; Bulletin of the Atomic Scientists, July-August, 1990, pp. 19-21.

Durch I. "International Arms Market" ; Defense News, March 14-20, 1994, p. 16

Dutta S. "Opportunities and Prospects for Indo-US. Cooperation on Asian Security Issues: China and South East Asia." In The United States and India in the Post-Soviet World; Proceedings of the Third Indo-US Strategic Symposium, Virginia, April 1992.

ECO - "Its Broke, So Fix it : The Nuclear Nonproliferation Treaty is in Urgent Need of Repair", The Economist, 27 July 1991, p.13.

Elmandjra M. "The Impacts of Social and Cultural Environment on the Development of
Information Technology"; Technology Transfer for Development -The Prospects and Limits of Information Technology. (Summary by Ansen D. and Kells K.); in Rueschlikon, Zurich, Switzerland. 5 December, 1991.

Erlich J. "U.S. Wins in War Games / Study says Smart Weapons Will Run Out Before Victory." In Air Force Times, 19 August 1996-A.

Erlich J. "Future of Multinational Export Control Remains in Question"; Defense News, 22-28 July 1996-B, pp. 10.

Erlich J. U.S. Lawmakers Say China's Sales Warrant Penalties; Defense News,14 April 1997, p3

FAS - "Arms Sales Code of Conduct - Arms Transfer Code of Conduct: Talking Points"; Arms Sales Monitoring Project, Federation of American Scientists, February 1995.

Ferrari P.L. US Arms Exports: Policies and Contractors; Cambridge, Mass. Ballinger, 1988.

Fialka J.J. "War By Other Means- Economic Espionage in America", W.W. Norton & Company, 1997

Finnegan P. "Middle East Focus Shifts to Upgrades"; Defense News, March 14-20, 1994, p. 12.

Fisher C.S. "The preconditions of confidence building-lessons from the European Experience"; in A Handbook of Confidence Building Measures for Regional Security, the Henry L. Stimpson Center, September 1993. pp. 31-45

Fisher R.D. "Why Asia is not ready for Arms Control." In Backgrounder - Asian Studies Center, The Heritage Foundation, No: 113, 25 May 1991.

Fisher R.D. 'How America's Friends Are Building China's Military Power', The Heritage Foundation, Roe Backgrounder No. 1146, 5 November, 1997

Fogleman R., "America's Air Power Team." Remarks to the National Security Industrial Association West Coast Dinner, Century City, Calif., Jan. 19, 1996.

Frank A.G. Crisis : In the Third World, Holmes & Meier Publishers, 1981 pp. 280-310.

Freeman C.W. The Diplomat's Dictionary, National Defense University Press, 1994. p.379.

Gansler J.S. - 2 "Defense Conversion - Transforming the Arsenal of Democracy"; MIT Press, 1995; p. 169.

Gansler J. S., "Constructively Transforming the Russian Defense Industry," February 28, 1994-A, pp. 16-17.

Gansler J.S. "Defense Conversion: Transforming the US Defense Industrial Base," Survival (Winter 1993-1994-B), pp. 130-46.

Gertz B. "US firms' tips boosted Chinese missile program"; The Washington Times, Tuesday,14 April 1998-A.

Gertz B. " China Targets Nukes at US ';The Washington Times, 1 May 1998 and 'China seen as security Threat', The Washington Times, 4 May 1998-B

Gibbons J. H. Foreword, "Holding the Edge: Maintaining the Defense Technology Base"; OTA-ISC-420. Office of Technology Assessment, Washington, DC: US Government Printing Office, April 1989.

GOI-DDPS   Indian Defense Products, Government of India, Department of Defense Production and Supplies, New Delhi-110011, India, 1994 .

GOI-MCOM   Ministry of Commerce, Public Notice No. 68 Exp (PN) / 92-97, Government of India, New Delhi, 31 March, 1995.

Goldring N. J. "Need weapons? Just ask Uncle Sam"; Chicago Tribune, 22 April 1996.

Gorbachev M. Perestroika - New thinking for Our Country and the World; Harper & Row, New York, 1987. pp. 185.

Graham B. 'Interceptor Test Cost To Pass $70 Million-Analysts Urge More Trials After Failures' ;Washington Post, April 27, 1998; Page A09

Graves, N. "Under the gun to build nukes, India stands firm on test ban" ; The Washington Post, 4 July 1996.

Grimmtt R.F. "Trends in Conventional Arms transfers to the Third World by major Supplier, 1982-1989."; Congressional Research Service, June 1990-A.

Haggard S. "The ACP States, the European Community Associates, and India." In Developing nations and the Politics of Global Integration ; The Brookings Institution, Washington D.C., 1993; pp. 100-101.

Haggard S. Developing Nations and the Politics of Global Integration; The Brookings Institute, Washington D.C., 1995.

Hamilton L.H. "Strengthening US-India Relations" in address to Asia Society, Washington DC, 29 April, 1994.

Haniffa A. "A conditional U.S. offer on technology'. In India Abroad, 27 March 1998,p23

Haniffa A. "India seen as structured to Catch up with China." In India Abroad, 26 July 1996-A.

Haniffa A. "China Reported Ahead in Military Technology." In India Abroad, pp22, 26 July 1996-B.

Haniffa A. Statement of Christopher Warren, In India Abroad, 16 Aug 1996-C

Haniffa A. "US Cites India's Restraint on Nuclear Weapons." India Abroad, Washington 26 April 1996-D.

Harrison S.S. "South Asia and the United States: A Chance for a Fresh Start." Current History, March 1992, Vol. 91, No. 563, pp. 97-105.

Harrison S.S. 'The United States and South Asia : Trapped by the Past" in Asia: South Asia at 50: A Retrospective; Current History, December 1997 C:\archivedec97\currentissue.html

Harrison S.S. and Kemp G. "India and America After the Cold War", Carnegie Endowment for International Peace, Washington D.C. 1993.

Hitchens T. "US drive to Cooperate may be slowed by Rules." In Defense News 8-14 April 1996.

Hua D., Morgan G. and Wulf H. "Export of Dual-Use Technology in Arms Sales Versus Nonproliferation: Economic and Political Considerations of Supply, Demand and Control:. In AAAS Science and Security Colloquium; 1991.

Hurewitz B.J. "Non-Proliferation and Free Access to Space: The Dual-Use Dilemma of the Outer Space Treaty and the Missile Technology Control Regime." High Technology Law Journal Abstract, Issue 9:2 - Spring 1994.

IDR "Developing International Transparency: Successes for the United Nations Register of Conventional Arms," International Defense Review, vol. 5; 1994, pp. 23-27.

IDR "New Sino-Pakistani MBT Project," International Defense Review, December 1988, p. 1553.

Jasjit Singh "Indian missiles versus US double standards"; The Statesman, New Delhi, 20 June 1996.

Jasjit Singh The NPT Debate and the CTBT Stand-Off; AGNI - Studies in International Issues, Vol 1, Number 1, April 1995.

Jha P.S. "Is a South Asian Free Trade Area Possible ? The Political Limits of Cooperation." In India and Its Neighborhood; Ed. US Bajpai; Lancer International Delhi, 1986. p. 150.

Johnson J. "Conventional Arms Transfer Policy - An Industry Perspective." Testimony Before a Joint Hearing of the Committee on Foreign Affairs ; Subcommittee on Arms Control, International Security and Science and the Subcommittee on Europe and the Middle East House of Representatives; 27 May, 1992.

Joshi M. " Dousing the Fire ? : Indian Missile Programme and the United States' Nonproliferation Policy ." In Strategic Analysis Vol XVII No. 5. August 1994, pp. 557-575.

Kalam A. A.P.J. " Technology as a vehicle of growth for a strong and vibrant India"; Technology Summit '95, New Delhi, 14 December 1995-A.

Kalam A.P.J. "Assistive Devices : Scope and Challenge Ahead." Ten Day Technical Workshop on Indigenous Production and Distribution of Assistive Devices; Economic and Societal Commission for Asia and Pacific Region (ESCAP) and Ministry of Welfare. Madras, India, 5 September, 1995-B.

Kapur A. "Dump the Treaty", Bulletin of the Atomic Scientists, July-August, 1990 pp. 21-23.

Karp A. "Trade in Conventional Weapons,." In SIPRI Yearbook 1988: World Armaments and Disarmament (Stockholm: SIPRI, 1988), p. 178.

Kemp G. "Proliferation on the Subcontinent Possibilities for US-Indian Cooperation": The United States and India in the Post-Soviet World; Proceedings of the Third Indo-US Strategic Symposium, Virginia,21-23,April 1992; Pub: National Defense University, Washington DC, 1993.

Khan M. A. 'Security Implications of Nuclear Proliferation in South Asia'; Weapons of Mass Destruction: New Perspectives on Counter proliferation, Edited by William H. Lewis and Stuart E. Johnson, Center for Counter proliferation Research, NDU Press,1995. P76

Kimball D. "Status of Comprehensive Test Ban Treaty Negotiations"; Physicians for Social Responsibility ; updated on the Internet - January 29, 1996.

Koch A. "Nuclear Testing in South Asia and the CTBT." In The Nuclear Proliferation Review, Volume 3, Number 3, Spring-Summer 1996. pp. 98-104.

Kopte S., Renner M. and Wilke P. "The Cost of Disarmament: Dismantlement of Weapons and the Disposal of Military Surplus." In Nonproliferation Review, Monterey Institute of International Studies, Winter 1996; pp. 33-45.

Kranti V. "China is less important to the US than India." In The Observer, 17 June 1996.

Krepon M. McCoy D., Rodolph M.C.J. A Handbook of Confidence-Building Measures for Regional Security, The Henry L. Stimson Center, Washington, DC: September 1993.

Kurosawa M. "The Nuclear Non Proliferation Regime Beyond 1995." In The defense Trade: Demand supply and control. Volume 3; Royal Institute of International Affairs, London, 1994.

Laurance E. "Transparency in Armaments." In Missile Monitor, International Missile Proliferation Project, Monterey Institute of International Studies; ISSN 1060-8273, 2 November 1992. pp. 4-9.

Lelyveld M.S. " US Sales of Torture Devices Blasted" in The Journal of Commerce, 25 June 1996.

Lin Chong-Pin "China's Nuclear Weapons Strategy', Lexington Books, 1988, p27

Long W. "Defining Strategic Exports in the 1990s:From Export Control to the Management of Technology Exchange." In Export Controls In Transition; ed. Gary K.
Bertsch and Steven Elliot-Gower, Duke University Press, 1992; pp. 128-147.

Lubman S. and Cole J. "Weapons Merchants Are Going Great Guns in Post-Cold War Era"; Wall Street Journal, January 28, 1994, pp. Al- A6.

Lumpe L. 'Arms Sales Code of Conduct Campaign', Arms Transfer Working Group, Internet address http://www.fas.org/asmp/atwg/code/index.html, 12 Jan 1998

Macke R.C. Prepared Statement of Admiral Richard C. Macke, US Navy, Commander in Chief, United States Pacific Command, before the House International Relations Committee Subcommittee on Asia and the Pacific ; 27 June 1995.

Maghroori R. and Ramberg B. "Globalism Versus Realism : A Reconciliation." In Globalism Versus Realism: International Relations Third Debate: Eds. Maghroori R and Ramberg B., Westview Press; 1982 ; pp. 223-231.

Mama, H India Offers Plastics; International Defense Review, March 1992, p. 44.

Martel W.C. and Pendley W.T. Rethinking US Proliferation Policy for the Future : Weapons of Mass Destruction : New Perspectives of Counter proliferation : Institute of National Strategic Studies. Ed. by William H. Lewis and Stuart E. Johnson. National Defense University Press, Washington D.C. April 1995.

McCarthy, C. "Global poverty and Weapons Giveaways"; The Washington Post, 9 July 1996.

Miller S.E. "Assistance to Newly Proliferating Nations." In New Nuclear Nations - Consequences for US Policy, Eds. Robert D. Blackwell and Albert carnesale, 1993; p. 113.

Milhollin G. "Should we sell Supercomputers to Algeria ?" ; The New York Times, 24 April 1998.

Morehead J.W. "Enforcing Export Controls: Improving the Effectiveness of US and Multilateral Export Controls." In Export Controls In Transition; ed. Gary K. Bertsch and Steven Elliot-Gower, Duke University Press, 1992 ; pp. 128-147.

Morgan M.G. "Arms Sales Versus Nonproliferation: Economic and Political considerations of supply, Demand and Control." AAAS Science and Security Colloquium, 1991.

Morocco J.D. "Arms Modernization Key Long-Term Goal." In Aviation Week and Space Technology, March 14, 1994, p. 48.

Morse J. 'US satisfied with China's nuclear controls'; US State Dept. Background Briefing, 31 October 1997

Myers S.L. "US Aides Describe How Russia helps India on Missiles"; New York Times, 27 April 1998, p1.

Najeeb M. "Missiles said to be capable of targeting Indian cities", India Abroad, 27 March 1998, p23.

Nau H. R. "Export Controls in a Changing Strategic Context." In Export Controls in Transition. Ed. Bertsch G. K and Gower S. E., Duke University Press, USA 1992.

Naughton P. Reuter News, Geneva 9 April 1996.

Nayyar K.K. "Emerging Areas of Conflict in the Twenty First Century - A Third World Perspective"; AGNI, Studies in International Strategic Issues; Volume 1, Issue 1, April 1995.

Nebehay S. Reuter News, Geneva 29 March 1996.

Nichols R.W. "Science, Technology and government for a Changing World" ; Carnegie Commission on Science, Technology and Government, April 1993.

Nolan J.E. "Toward an International Technology Security, Regime." In Trappings of Power: Ballistic Missiles in the Third World; Brookings Institution, 1991.

Nolan J. E. 1994-A "The Imperatives for Cooperation, Global Engagement"; Cooperation and Security in the 21st Century ; Brookings Institution, 1994. p. 27.

Nolan J. E. "Preventive Approaches : The MTCR Regime." In Weapons of Mass Destruction : New Perspectives of Counter proliferation : Institute of National Strategic Studies. Ed. William H. Lewis and Stuart E. Johnson. National Defense University Press, Washington D.C. April 1995.

Nunn S. 'US Policy Toward China'; Russell Symposium, 1997 Proceedings, University of Georgia, Athens, Georgia, 3 November 1997

O'Hanlon M. "Defense Planning for the Late 1990s"; Studies in Defense Policy, The Brookings Institution, 1995.

Opall B. 'White House Resists Tighter Rein On Supercomputer Sales / Experts Say Exports To China Pose National Security Risk', Defense News, 28 April 1997; p6

Opall B. 'China Pact Eases Grip on Nuclear Licenses'; Defense News, 4-10 May 1998, p4

O'Prey K. "The Arms Export Challenge; Cooperative Approaches to Export Management and Defense Conversion" ; Brookings Occasional Papers, The Brookings Institution,1995.

Ostry S. and Nelson R.R. Techno-Nationalism and Techno-Globalism - Conflict and Cooperation; The Brookings Institution, 1995.

OTA "Technologies underlying Weapons of Mass Destruction." Office of Technology Assessment, December 1993; p. 222.

OTA "Assessing the Potential for Civil-Military Technology, Processes, and Practices"; US Congress, Office of Technology Assessment; OTA-ISS-611 ;Washington, D.C., US Government Printing Office, September 1994.

Ozga D. A. "Chronology of the Missile Technology Control Regime." Center for Nonproliferation Studies; The Nonproliferation Review: Monterey Institute of International Studies; Volume 1 - Number 2. Winter 1994.

Paige E. Jr. 'Retaining the Edge on Current and Future Battlefields' : Defense Issues: Volume 10, Number 85, Aug. 22, 1995. Prepared remarks by Emmett Paige Jr., assistant secretary of defense for command, control, communications and intelligence, to the Fort Bragg chapter of the Armed Forces Communications and Electronics Association, Fayetteville, N.C.

Perkovich G. "Three Models for Nuclear Policy in South Asia: The Case for Non-weaponized Deterrence." In Weapons of Mass Destruction : New Perspectives on Counter proliferation; National Defense University Press, April 1995.

Perry W.J. "Ever Vigilant in the Asia-Pacific Region", Defense Issues: Volume 10, Number 87; Remarks by Secretary of Defense

William J. Perry to the Japan Society, New York City, Sept. 12, 1995-A.

Perry W.J. "The Ingredients for Democracy." Defense Issues: Opening remarks at the Defense Ministerial of the Americas, Williamsburg, Va. Volume 10, Number 86, July 25, 1995-B.

Pincus W. "New methods Help Maintain Nuclear Arms", Washington Post, 28 April 1998, pA6.

Pollard R. "Some Issues in Technology Transfer"; Draft document for the Technology Transfer Task Group of the NGO Strategy Group for UNCED, at 2nd UNCED PrepCom II, Geneva, April 1991.

Preston C. "DOD Must Re-engineer Its Procurement System Now." Prepared statement of, deputy undersecretary of defense for acquisition reform, to the House Government Reform and Oversight Committee. : Defense Issues: Volume 10, Number 24-- Feb. 21, 1995.

Priest D. "US Goes Easy on Allies in Arms Control Crusade"; Washington Post, 14 April 1998, pA1,A11.

Radyuhin V. 'Moscow stands by India on Kashmir'; The Hindu, 16 March 1998

Raghuvanshi V. "India Strives To Cut Import Reliance by Easing Limits." In Defense News, October 16-22, 1995, p. 70.

Raghuvanshi V. "New Delhi's Defense Budget Leaves Little Room To Grow." In Defense News, August 5-11, 1996; p. 9.

Raja Mohan C. "US Unhappy about Indian Position on CTBT." The Hindu, 27 November 1995.

Rajghatta, C "US ducks law, Plan technology sale to China." In Indian Express, New Delhi, 21 June 1996.

Rees-Mogg W. "In praise of India." In The Times of London; 11 March 1996.

Reinicke W.H. "From Denial To Disclosure : The Political Economy of Export Controls and Technology Transfer." Joint Indo-American Seminar on Non-Proliferation and Technology Transfer, University of Pennsylvania October 3-6, 1993.

Reinsch W.A. Bureau of Export Administration; Department of Commerce, US., 1995.

Reuter, New Delhi, India. Posted on the Internet at 12:34, 30 March, 1996.

Risdon, Penny, 1992. "Understanding The Technology Transfer Process." Vita Distribution Service, 20 Dec 1992.

Risk Report Referenced here - Published ten times a year by the Wisconsin Project on Nuclear Arms Control. ISSN 1080-2916, 1701 K. Street, NW, Suite 805, Washington DC-1995.

Rodan S. And Raghuvanshi V. 'Israel, India cooperation accord may trigger sales'; Defense News, 21 October 1996,p27

Rosecrance R., Alexandroff A., Koehler W., Kroll J., Lacqueur S. and Stocker J. "Whither Interdependence?." In Globalism Versus Realism: International Relations Third Debate: Eds. Maghroori R and Ramberg B., Westview Press; 1982.

Sadowski Y. M. "Scuds or Butter-The Political economy of Arms Control in the Middle East ; the Brookings Institute, Washington D.C. 1993.

Sanders R. "India and Israel in Arms Industries : New Suppliers and Regional Security"; National Defense University, Washington D.C. 1990.

Santhanam K. "Opportunities and Prospects for Indo-US Cooperation in Defense Technologies"; In The United States and India in the Post-Soviet World; Proceedings of the Third Indo-US Strategic Symposium, Virginia, April 1992.

SECC "The Future of Security Export Controls" (Unofficial translation) Security Export Control Committee (SECC), industrial structure Council, Ministry of International Trade and Industry, 25 March 1993.

Shalikashvili J.M. "Grand Challenges for the Post-Cold War World." Defense Issues: Volume 10, Number 69 Remarks as delivered at the Naval War College graduation, Newport, R.I., June 16, 1995.

Sharma A 'India Conducts Three Nuclear Tests', Associated Press Writer, Washingtonpost.com:World News, 11 May 1998

Singh R.P. "US Policymaking on Military Technology Controls: A Case Study Missile Technology Control Regime." Advanced Seminar on the Foreign Policy Process, University of Maryland, July 1990.

Singh S. K. "Nuclear Weapon Free World : Prospects." In AGNI - Studies in International Strategic Issues, Vol 1, Number 1, April 1995.

Smith R.J. US "Aides See Troubling Trend In China - Pakistan Nuclear Ties." In Washington Post, 1 April 1996.

Smith J.G "India and the Bomb - Public Opinion and Nuclear Options"; eds. David Cortright and Amitabh Mattoo; University of Notre Dame, Notre Dame, Indiana. 1996 ; pp. 109-34.

Smith B.L.R. and Barfield C.E. Technology. R&D and the Economy, The Brookings Institution and American Enterprise Institute, Washington D.C., 1996 (pp75-139)

Snyder J. C. After the Cold War - South Asian Security The US Approach to the Indian Sub-Continent Strategic Forum 43 . August, 1995.

Steinbruner J. and Nolan J. Testimony before the Subcommittee on Economic Policy, Trade, and Environment of the House Committee on Foreign Affairs, on June 9, 1993 and June 23, 1993.

Steinbruner J.D. "Weapons Production and Force Structure Planning as a Problem of Collaboration," International Arms Sales Conference held by the Brookings Institution and the Institute for USA and Canada Studies, Queenstown, Md., February 3-5. 1994, p. 2.

Subrahmanyam K. " Test Ban Seen as Perpetuating Nuclear Apartheid." In India Abroad, 16 August, 1996, p. 2.

Subrahmanyam K. "India and its Neighbors : A Conceptual Framework of Peaceful Co-existence." In India and Its Neighborhood; Ed. US Bajpai; Lancer International Delhi, 1986. pp. 109-139.

Subrahmanyam K. "Opportunities for Indo-US Cooperation on Arms Control and Non Proliferation." In The United States and India in the Post-Soviet World; Proceedings of the Third Indo-US Strategic Symposium, Virginia, April 1992.

Tahir-Kheli S.R. 'India, Pakistan and the United States-Breaking With the Past', Council on Foreign Relations Press, 1997

Tanham G.K. "Indian Strategic Thought - An Interpretive Essay." R-4207-USDP, RAND, Santa Monica, CA, USA, 1992.

Taylor T. "Conventional Arms : The Drives to Export" ; The Defense Trade: Demand supply and control. Volume 3; Royal Institute of International Affairs, London, 1994.

TIUSS : The United States and India in the Post-Soviet World : Proceedings of the Third Indo-US Strategic Symposium, Virginia, April 1992. p3

Turner S : Caging the Nuclear Genie – An American Challenge for Global Security, Westview Press,1997

UNISPACE National paper of India for the Second UN conference on exploration and peaceful uses of outer space (UNISPACE 82); Government of India, Publications and Public Relations Unit, ISRO, 1981, Bangalore, p.36.

Velisarios Kattoulas Report by Reuter, Beijing, 13 June 1996

Victor C. India : The Security Dilemma. New Delhi: Patriot Publishers, 1990; p. 62.

Wallop M. and Codevilla A. The Arms Control Delusion; ICS press, Institute of Contemporary Studies, San Francisco, California, 1987.

Washburn J. "Arming the World to the teeth." In The Sun, 23 July 1996.

Watanabe O. "Major Changes in Global Security Controls and the Role of Japan"; The defense Trade: Demand supply and control. Volume 3; Royal Institute of International Affairs, London, 1994.

Weiner T. "US Says North Korea Helped Develop New Pakistani Missiles"; The New York Times, 11 April 1998

WH The White House; "Conventional Arms Transfer Policy", Office of the Press Secretary; Fact Sheet: February 17, 1995.

White J. "The Compelling Case for Modernization"; Remarks by Deputy Defense Secretary to the Air Force Association Business Session, Washington : Defense Issues: Volume 10, Number 89. Sept. 18, 1995.

Widnall S.E. "Crossflow of Technology Across International Bounds." Remarks to the US Chamber of Commerce, Canberra, Australia, Oct. 19, 1995 (A).

Widnall S.E. "Which Way to the Revolution?." Remarks to the Air Force Association National Symposium in Los Angeles, Oct. 27, 1995-C.

Widnall S.E. 'The Choice and the Opportunity Are Ours." In Defense Issues: Volume 10, Number 70 ; Remarks to the National Security Forum, Maxwell Air Force Base, Ala. June 2, 1995-B,

WT   Washington Times, 25 July 1989 p A2.

WWCAT " World Wide Conventional Arms Trade (1994-2000) - A Forecast and Analysis." US Department of Defense, December 1994.

WWR-1990   World Weapons Review, 27 June 1990, p. 2.

Zimmermann T., Chellaney B. and Moulier P.B. "Nuclear face-off-. India and Pakistan raise the stakes" in US News, 12 February, 1996.

R. J. Augustus

 # Glossary Of Acronyms

ACDA – Arms Control and Disarmament Agency

ARF – Asian Regional Forum

AG - Australia Group

ANWFZ - African Nuclear-Weapon-Free Zone

ASEAN - Association of Southeast Asian Nations

BARC – Bhaba Atomic Research Centre

BW - Biological Weapons

BWC - Biological Weapons Convention

CAS - Committee on Assurances of Supply (IAEA)

CBM – Confidence Building Measures

CBWNP - Chemical and Biological Weapons Nonproliferation Project

CD - Conference on Disarmament

COCOM - Coordinating Committee for Multilateral Export Controls

COTS – Commercial Off the Shelf (Technology)

CTBT - Comprehensive Test Ban Treaty

CW - Chemical Weapons

CWC - Chemical Weapons Convention

DCS – Direct Commercial Sales

DODD – Department of Defense Directive

DPG – Defense Policy Group

DRDO – Defence Research and Development Organization

DTSA – Defense Technology Security Administration

ESARDA - European Safeguards Research and Development Association

EU - European Union

EURATOM - European Atomic Energy Community

FMS – Foreign Military Sales

GOI – Government of India

IAEA - International Atomic Energy Agency

IAEL - International Atomic Energy List (COCOM)

IIL - International Industrial List (COCOM)

IML - International Munitions List (COCOM)

ISTCs - International Science and Technology Centers

JTG – Joint Technical Group

LOA – Letter of Offer and Acceptance

MOD – Ministry of Defence (India)

MTCR - Missile Technology Control Regime

NPT - Treaty on the Non-Proliferation of Nuclear Weapons

NSG - Nuclear Suppliers Group

NWS - Nuclear-Weapon States

NWFZ - Nuclear Weapon-Free Zone

NWSs - Nuclear-Weapon States

OAU - Organization of African Unity

OECD - Organization for Economic Cooperation and Development

OPANAL - Agency for the Prohibition of Nuclear Weapons in Latin America

OPCW - Organization for the Prohibition of Chemical Weapons

OSCE - Organization for Security and Cooperation in Europe

OST – Outer Space Treaty

PTBT - Partial Test Ban Treaty

SAARC - South Asian Association for Regional Cooperation

R. J. Augustus

SAGSI - Standing Advisory Group on Safeguards Implementation (IAEA)

SEANFZ - Southeast Asia Nuclear-Free Zone (ASEAN)

SPNFZ - South Pacific Nuclear-Free Zone

UN - United Nations

UNDC - United Nations Disarmament Commission

UNGA - United Nations General Assembly

UNIDIR - United Nations Institute for Disarmament Research

UNSC - United Nations Security Council

UNSCOM - United Nations Special Commission on Iraq

USG – United States Government

VEREX - Ad Hoc Group of Government Experts (BWC)

ZOPFAN - Zone of Peace, Freedom and Neutrality (ASEAN)

In Defense Of My Country

R. J. Augustus

Index

A

academia, 3, 153, 169, 201
accessible, 11
adversaries, 5, 50, 107, 144, 198
advocate, xiii, 37, 50, 66
aerospace, 63, 103, 160, 227
Aerospace Industries Association, 61, 63, 284
affordable, 11, 181, 184, 249
AFPD, 139
agenda, 18, 51, 240, 262, 264
Agni, 76, 86, 89, 114, 187, 261
Agreed Minutes', 235
agreement, 16
agreements, 46, 47, 50, 56, 67, 107, 134, 142, 152, 153, 154, 167, 174, 214, 215, 217, 223, 237
Ahearn, 43, 283
AIA, 61, 210, 284
Akash, 228
Albright, 124, 284
Ali, 23, 261, 284
allies, 3, 5, 10, 11, 17, 39, 47, 55, 57, 77, 94, 96, 98, 103, 107, 121, 144, 179, 227, 256
ambitions, 40, 42, 77, 78, 191, 194, 263
American Technology Preeminence Act, 142
Anatomy, iii, 29
animosities, 81, 100
annihilation, 99
anxiety, 53, 76, 159
Apotekar, 121, 284
apparatus, 163
Appendix, x, 224
Application Phase, 151
Arab, 48
Argentina, 34, 50, 115

armed forces, 13, 33, 52, 55, 63, 87, 179, 187, 202, 205, 207, 232, 233, 241, 249
arms race, 25, 26, 115, 128, 209, 216, 232
Arms Race, 23, 287
Arms Transfers, iii, 50
arms-length, 46
Arnet, 183
array, 97, 126
Arrow, 101
arsenal, 45, 48, 116, 205
articles of agreement, 157, 158
Arunachalam, xv, 129, 252
Asia, 18
Asian, iv, 19, 26, 31, 33, 38, 40, 41, 42, 43, 63, 71, 74, 87, 125, 256, 284, 288, 289, 290, 294, 303, 307, 309
assurances, 78, 224, 276, 277, 278, 279
atmospheric nuclear tests, 117
attractive, 20, 45, 59, 100, 103, 181
Australia Group, 122, 123, 124, 247, 307

B

balance, 40, 41, 87, 93, 195
balanced strategy, 39
ballistic missiles, 27, 72, 79, 112, 114, 189, 219
BARC, 130
Barnard, 45, 285
Barnes, 134, 285
barrier, 114
barriers, xi, 38, 83, 126, 135, 147, 167, 168, 169, 170, 180, 197, 200, 264
barriers to technology transfer, 135, 169, 264
Basu, 120
battery, 101
battlefield, 20, 22, 124, 179, 206

313

battlespace, 20, 194
Bayh-Dole Act, 142
be confidence building measures, 100
behavior, 37, 64, 122, 151, 169, 186, 239
BEL, 130
Bernstein, 58, 285
Bhaba Atomic Research Center, 130
Bharat Electronics Ltd, 130
Bhutan, 74, 125
Bhutto, 23
Bidwai, 120, 285
bilateral, xiii, 2, 7, 18, 29, 44, 133, 134, 174, 231, 236, 264
biological, 65, 96, 110, 124, 128
Biological Arms Control Institute, 94
Biological Weapons Convention, 122, 307
bipolar, 9
Blackwell, 13
blue ribbon panel, 134
Bofors, 50
border, 26, 38, 42, 74, 75, 77, 80, 86, 136
Boyer, 65, 285
bracketing, 44
Brazil, 34, 67, 114, 115, 283
Britain, 30, 50, 60, 77, 118
Brookings Institution, 208, 283, 286, 292, 298, 299, 303
Bryen, 60
Bureau of Export Administration, 171, 172, 286, 301
bureaucracies, 35, 135, 239, 240
bureaucrats, 22, 58
business, 15, 25, 38, 41, 45, 46, 49, 54, 58, 59, 74, 77, 96, 102, 108, 136, 137, 150, 161, 184, 190, 196, 197, 198, 200, 201, 203, 211, 212, 217, 220, 221, 238, 249, 250, 256
BXA, 171, 172, 174, 286

C

capabilities, 14, 17, 18, 20, 21, 22, 26, 33, 35, 46, 55, 66, 76, 78, 96, 103, 110, 112, 116, 120, 153, 160, 161, 169, 179, 180, 184, 185, 189, 194, 198, 199, 202, 216, 218, 228, 232, 262
Cardomone, 65, 286
Carnesale, 13
Carr, 159, 286
Carter, 97, 98, 211, 262, 286
Caruana, 200, 286
CAVCTS, 186
chain reaction, 101
challenges, 9
Chalmers, 232, 286
changes, 12
Charlie Rose, 43
Chellaney, 186, 285, 305
chemical, 65, 79, 110, 111, 124, 125, 128
China, 16, 18, 25, 26, 31, 34, 36, 40, 43, 48, 53, 60, 64, 71, 73, 74, 76, 77, 78, 80, 83, 84, 85, 86, 87, 89, 90, 100, 112, 115, 118, 119, 120, 121, 127, 130, 189, 194, 233, 260, 261, 263, 265, 287, 288, 289, 290, 291, 292, 295, 296, 297, 298, 301, 302
Chinese, 18, 22, 26, 34, 36, 53, 74, 76, 78, 86, 87, 154, 258, 291
Christopher, 120, 292
CIA, 23, 25, 76
civilian, xi, 20, 57, 63, 64, 80, 112, 126, 177, 182, 187, 199, 202, 228, 235, 247, 251, 255, 256, 260
clandestine, 111, 115, 124
Clinton, 13, 27, 34, 77, 78, 120, 141, 174, 206, 235, 284
Clinton's, 141
CLSSA, 221
COCOM, 6, 121, 122, 126, 307, 308
Code of Conduct, 54, 218, 290, 296
code of ethics, 154
Codevilla, 49, 304
coercive diplomacy, 46
Cohen, 101
Cold War, 2, 5, 11, 16, 18, 23, 29, 32, 39, 57, 73, 100, 108, 121, 122, 172, 173, 208, 231, 232, 233, 285, 287, 293, 296, 302, 303
Coll, 239, 287
collaboration, 38, 74, 125, 133, 134, 238, 263
collapse, 9

combat, 22, 56, 66, 125, 160, 178, 183, 193, 252, 262
commanders, 22
commercial, 3, 33, 46, 52, 57, 98, 102, 103, 124, 131, 133, 136, 139, 141, 142, 148, 150, 168, 173, 177, 178, 181, 182, 183, 184, 187, 188, 196, 197, 198, 199, 201, 202, 211, 220, 221, 228, 238, 248, 249, 257, 264
Commercial Phase, 150
commodities, 7, 57, 62, 108, 144, 145, 223
common faith, 29
common ground, xiii, 7, 33, 244
common ground., 7
community, 3, 10, 31, 36, 53, 113, 115, 116, 118, 120, 199, 201, 238
competitive, 4, 58, 67, 96, 97, 102, 152, 171, 172, 175, 190, 221, 228, 238, 251, 260
compromise, 61, 216
conceptual, 148, 149
Conceptual Phase, 149
concerns, xi, xii, xiii, 2, 5, 17, 19, 22, 31, 32, 41, 55, 60, 82, 96, 109, 115, 120, 123, 129, 134, 151, 154, 173, 175, 194, 203, 239, 240, 244, 246, 247, 257, 258, 263
conclusion, 11
conclusions, xiii, 26
Conference on Disarmament, 118, 129, 307
conferences, 154
Confidence Building Measures, 231, 290
conflicts, 5, 11, 12, 19, 73, 98, 100, 106, 154, 173, 183, 203
confusion, 5, 7, 75, 124, 239
congruence, 12
consensus, 6, 16, 39, 81, 86, 97, 122, 208, 231
consistent export policy, 61
constraints, xii, 2, 21, 38, 120, 179
contractors, 185, 216
conventional, xii, 5, 17, 25, 49, 51, 65, 85, 93, 95, 96, 97, 105, 109, 111, 117, 122, 156, 208, 216, 232, 240, 256, 263, 285, 288

Conventional Arms, 54, 55, 105, 121, 135, 144, 218, 286, 292, 293, 294, 304, 305
conversion, 53, 57, 58, 136, 173
Cooper, 23, 46, 287
cooperation, 12
Cooperative Research and Development Agreements, 142, 152
Core Technologies, v, 181
cornerstone, 101, 258
corruption, 50
cost, 4
cost overruns, 131, 197
cost-effective, 5
COTS, 98, 182
CRDA, 152
crown jewels, 3
crown jewels', 141
Cryogenic, 110, 114
CTBT, 7, 48, 116, 117, 118, 119, 120, 121, 129, 130, 134, 244, 285, 293, 295, 300, 308
culture, 1, 36, 41, 71, 82, 161, 248
customer, 45, 62, 149, 151, 201, 210, 222
cutting edge, 3, 102, 126, 202
Cyranek, 168, 287
Czerwinski, 10

D

data, 24, 52, 79, 138, 150, 182, 212
deadlock, 61
deal, xi, 3, 50, 59, 60, 73, 85, 110, 114, 157, 159, 173, 174, 197, 221, 239, 240, 247, 261, 264, 265
defense budget, 14
defense budgets, 4
Defense News, 61, 63, 122, 285, 287, 289, 290, 293, 298, 300, 301
Defense Policy Group, 236, 256
defense services, 19, 22, 153, 163, 190, 195, 196, 198, 201, 210, 235
defense spending, 15, 33, 58, 69, 84, 88, 101
defense technology, xi, xii, xiii, 2, 4, 7, 13, 15, 19, 21, 32, 44, 46, 49, 51, 55, 56, 57, 64, 65, 75, 87, 98, 101, 103, 135, 145, 153, 178, 179, 183, 184,

190, 195, 205, 206, 210, 234, 239, 241, 243, 246, 249, 251, 254, 257, 263, 264, 265
Defense Technology Plan, 102
delivery platforms, 109, 111
demarcation, 148, 256
democracies, 29, 30, 45, 233
democracy, 30, 32, 34, 36, 39, 41, 73, 82, 234, 255, 259
denial, 49, 131, 157, 171, 174, 186, 216, 245, 247, 256
Denman, 201, 287
depended, 14
dependency, 14, 46, 65, 167
deprive, 49
derived, 51, 101
Deshingar, 87
Deshpande, 86, 288
deterrent, 24, 194, 233, 246
detractors, 124
Deutch, 102
develop, 16
developed countries, 3, 21, 128, 141, 190, 198, 199, 223, 252, 253, 255
dialogue, 47, 86, 131, 240
dignity, 43, 49
dilemma, 49, 62, 257
diminish, 13
Direct Commercial Sales, 147, 219, 288
Dirksen, 72
DISAM, 220, 288
disconcerting, 94
discriminatory, xii, 113, 115, 116, 120, 128, 247
disguised, 99
distrust, 133
DOD Directive, 143
dogmas, 15
domestic, 18, 41, 43, 53, 59, 65, 66, 68, 73, 142, 156, 210, 213, 219, 265
dominance, 32, 178
Donowaki, 105, 288
Dorn, 206, 288
DPG, 236
Drace, 106, 289
DRDO, v, xiv, xv, 2, 7, 62, 90, 112, 114, 131, 153, 155, 156, 163, 166, 167, 178, 181, 183, 187, 188, 190,

192, 196, 197, 198, 199, 200, 201, 202, 203, 217, 226, 227, 248, 249, 250, 251, 252, 253
driving forces, 105
DTSA, 60, 143
dual use, xi, 47, 109, 110, 111, 121, 131, 171, 179, 182, 199, 247, 257, 258
Durch, 66, 289
dynamic, 42, 181, 200

E

EAR, 172
economic growth, 3, 34, 42, 77, 192, 262
Egypt, 94, 118
eligibility, 54
Elmandjra, 143, 289
emergent missile power, 20
endeavor, 22, 39, 42, 103, 144
enemy, 23, 32, 40, 65, 103, 106, 111
engagement, 13
engine, 39, 42, 65, 110, 186
enhanced, 16
environment, xiii, 5, 7, 13, 15, 18, 19, 22, 39, 58, 63, 100, 103, 108, 137, 147, 150, 151, 158, 168, 172, 173, 183, 197, 202, 203, 206, 207, 248, 253, 254
equality, xiii, 32, 43, 117
equipment, 20, 33, 35, 46, 48, 49, 50, 51, 55, 56, 58, 59, 62, 64, 67, 69, 70, 78, 106, 108, 110, 111, 121, 123, 124, 126, 138, 147, 152, 160, 161, 171, 177, 182, 185, 187, 188, 189, 195, 205, 208, 210, 217, 219, 222, 225, 227, 232, 251, 253, 254, 257
equipping, 11
espionage, 3
establishments, 2, 7, 21, 62, 95, 103, 131, 138, 153, 159, 177, 189, 190, 200, 202, 231, 233, 244, 250, 251, 253
evaluation, 4, 51, 147, 148, 151, 159, 166, 199, 234, 248
Evaluation Phase, 151
Executive Orders, 142
expenditure, 14, 59, 68, 69, 88, 90, 227

expertise, 20, 69, 96, 154, 161, 198, 225, 251, 258, 263
Export Administration Regulations, 172, 272
export controls, 5
export regimes, 6
exporters, 25, 47, 60, 65, 67, 108, 212
exports, 7, 44, 51, 52, 54, 55, 56, 57, 58, 59, 60, 61, 63, 66, 84, 87, 107, 121, 122, 123, 171, 174, 206, 207, 209, 210, 213, 215, 218, 227, 232, 243, 244, 247, 259, 261, 266

F

fear, xii, 23, 63, 73, 79, 96, 97, 98, 108, 124, 173, 190, 205, 232
fears, xii, 23, 70, 73, 95, 104, 119, 173
feasibility, 148, 150, 163
Feasibility Phase, 150
Federal Technology Transfer Act, 142
Fernandes, 119
fielded, 56, 136, 192, 197, 199
Finnegan, 67, 290
Fisher, 77, 212, 231, 290
flash points, 2
Fogleman, 20, 290
forecast, 54, 163
foreign exchange, 55, 56, 206, 227, 228
Foreign Military Sales, v, vii, ix, 147, 219, 220, 288
foreign policy, 14, 41, 53, 54, 55, 64, 65, 83, 124, 127, 133, 144, 171, 173, 185, 187, 193, 216, 218, 239, 246, 263
foundation, 11
framework, 13
France, 50, 60, 77, 101, 114, 115, 118, 122
Frank, 99, 290
frayed, 49
friendship, 35
functioning, 13
fundamentalism, 48, 80, 116
funding, 56, 90, 153, 183, 184, 252

G

Gansler, 39, 58, 202, 210, 291

Garten, 71
Germany, 60, 127
Gertz, 75, 77, 258, 291
Ghauri, 18, 25, 80, 86
Ghaznavi, 19
Gibbons, 101, 291
global influence, 40
global interests, 36
global policeman, 29
globalism, 37, 38
Goldring, 53, 54, 291
goodwill, 67
Gorbachev, 32, 291
Graves, 120, 292
GSOMIA, 134, 236, 244
guile, 45

H

Hamilton, 43, 292
Haniffa, 75, 87, 120, 183, 188, 219, 292
hardware, 49, 78, 84, 87, 138, 156, 188, 192, 213, 228
harmony, 37, 40, 44, 188, 231, 255
Harrington, 203
Harrison, 36, 245, 259, 292, 293
Hartman, 213
helicopters, 54, 189
high-tech, 38, 53, 103, 185, 187, 194, 248
history, xiii, 1, 16, 19, 29, 41, 45, 72, 81, 82, 83, 185, 194, 202, 260
horizontal proliferation, 129
hub, 42
human rights, 32, 45, 54, 216, 218
Hurewitz, 111, 113, 293

I

IAEA, 115, 307, 308, 310
ICBM, 265
IDR, 54, 232, 293
IGMDP, 186, 194, 228
illegality, 27
illegally, 51
illusion, 143
immigration of technology, 136
impediments, 37, 236

Import Certificate, viii, x, 224, 271, 273, 274, 275, 276, 278
in 'Operation Desert Storm, 39
incentives, 36, 45, 52, 227, 244, 266
independence, 30, 74, 117, 168, 185, 233
independent, 14
India Today, 119
Indian defense, 6, 13, 21, 24, 44, 45, 69, 87, 88, 89, 90, 183, 184, 188, 189, 194, 195, 196, 198, 199, 201, 203, 205, 211, 216, 259
indicators, 148, 150
indigenous, 2, 47, 84, 87, 107, 127, 131, 135, 186, 187, 189, 190, 194, 217, 249, 253
Indira Gandhi, 89, 130, 134, 185
Indonesia, 34
industrial base, 43, 45, 139, 182, 183, 189, 196, 199, 202, 205, 210, 211, 249, 255, 257
influence, 2, 3, 11, 31, 32, 37, 41, 51, 64, 66, 72, 80, 86, 98, 113, 114, 127, 139, 145, 150, 194, 217, 247, 259
information, 10
information warfare, 20, 96
innovation, 58, 139, 140, 177, 178, 202, 252
inroads, 101
instability, 10, 15, 23, 63, 100, 108, 240
integration, 24, 43, 82, 104, 151, 166, 199, 203, 211
Intellectual property, 138
interaction, v, xiii, xv, 3, 13, 29, 39, 43, 51, 54, 133, 137, 159, 161, 166, 167, 178, 189, 197, 199, 224, 234, 240, 263
interdependence, 10, 35, 37, 38
interests, 10, 15, 17, 29, 36, 40, 41, 43, 53, 55, 80, 88, 97, 104, 125, 128, 144, 171, 212, 231, 239, 246, 247, 248, 261, 262, 263
internal, 1, 3, 22, 40, 64, 72, 82, 137, 186
International Court, 27
international relations, 15
International Technology Transfer, xi, 148, 283

investment, 44, 45, 64, 71, 98, 155, 167, 180, 192, 198, 200, 228
Islamic, 48, 80
isolate, 40
isolating, 12
Israel, 17
ISRO, 110, 186, 228, 304
Italy, 60

J

Japan, 19, 34, 40, 85, 99, 127, 239, 265, 283, 300, 304
Jasjit Singh, 121, 188, 293
Jawan', 183
Jefferson', 265
Johnson, 207, 212, 217, 294, 296, 298
joint capability, 22
joint operations, 22
Joint Technical Group, 236, 256
jointness, 14
Joshi, 111, 114, 294
JTG, 236, 238
judge, 17
judgment, 17
juggernaut, 253
justification, 57, 79, 196, 211

K

Kalam, xiv, 187, 190, 194, 203, 294
Kaminski, 11, 195, 197
Karp, 87, 112, 294
Kemp, 35, 78, 84, 87, 113, 116, 126, 239, 260, 293, 294
Kennedy, 23, 234, 255
Khan, 31, 80
Koch, 130, 295
Kopte, 108, 295
Kranti, 43, 295
Krepon, 119, 231, 295

L

labyrinth, 143, 174
Lakshya', 228
launch, 86, 95, 154, 186, 228, 246
Laurance, 232, 295

LCA, 47, 201, 234, 252
lead time, 197
legislation, 79, 141, 142
legitimate, 3, 4, 22, 40, 48, 65, 66, 105, 109, 128, 144, 217, 228, 239, 246, 256, 257, 263, 266
Lelyveld, 59, 295
lesson, 15, 22
letters of credit, 157
leverage, 107, 249, 251, 255, 261
license, 51, 59, 93, 106, 130, 131, 136, 142, 143, 157, 171, 174, 175, 185, 189, 220, 225, 271, 276, 277, 278, 279, 281
Light Combat Aircraft, 47, 161, 201, 227, 234
Long, 19, 249, 255, 287, 296, 297
Lubman, 213, 296
Lumpe, 55, 296
lynch pin, 41

M

M-11, 26, 79
Macke, 235, 296
Maghroori, 37, 296, 301
Mahatma Gandhi, 30
Malaysia, 34, 56, 69, 227
Maldives, 74, 125
Mama, 227, 296
manhood, 24
manifest, 13
Market pull, 140
Martel, 16, 115, 296
Martin Luther King, 31
McCarthy, 213, 297
McDonald, 261, 284
measurable, 161
mechanism, 122, 125, 147, 160
mechanisms, 114, 122, 124, 152, 154, 246, 264
medicine, 41
Mexico, 34, 85
MiG-29, 56
Milhollin, 94, 297
military power, 5, 39, 55, 98, 194
military presence, 99, 100
military technology, 46, 47, 50, 77, 127, 144

missile, 18, 20, 24, 25, 26, 47, 75, 76, 78, 79, 80, 86, 95, 101, 103, 107, 110, 111, 112, 113, 120, 126, 127, 130, 134, 178, 185, 186, 187, 193, 194, 203, 219, 225, 228, 236, 239, 244, 246, 252, 260, 261, 291
missile technologies, 20, 227
Mission Areas, 234
MOD, 7
Modernization, 22, 297, 305
modernize, iii, 2, 20, 22, 46
Monterey Institute of International Studies, 20, 295, 299
Moodie, 94
Morehead, 50, 297
Morgan, 55, 293, 297
Moscow, 56, 80, 115, 300
mothballing, 108
motivations, 21
motives, 55, 75, 124, 173, 213
MOU, viii, 223, 224, 234, 271, 274, 278
MTCR, 7
MTOPS, 60
multilateral, 6, 34, 61, 100, 174, 208, 219, 231, 244, 256
multinational firms, 38
multinationals, 161, 262
multi-polar, 39
Muroyama, 173
myth, 143

N

naive, 265
Narasimha Rao, 89, 235
National Competitiveness Technology Transfer Act, 142
national security, 11
nationalism, 9
nationalism', 9
NATO, 47, 59, 69, 121, 239
Nayyar, 211, 298
Nebehay, 120, 298
negotiations, 39, 43, 59, 81, 111, 118, 129, 134, 157, 162, 166, 187, 220
neighborhood, 2, 11, 74, 85, 112, 194, 240
Nelson, 184, 283, 299

Nepal, 74, 125, 227
new thinking, 19
new world order, 12
Nichols, 39, 298
Nolan, 86, 105, 113, 208, 209, 298, 303
non-proliferation, xii, xiii, 33, 44, 53, 78, 109, 114, 116, 122, 246, 247
NPT, 7
Nuclear Club, 34
nuclear forces, 5
nuclear options, 17
Nuclear ownership, 17
nuclear program, 24, 76, 80, 119, 182, 259, 284
nuclear programs, 26, 78, 115
nuclear proliferation, 16
nuclear restraint, 33
Nuclear Suppliers Group, 112, 122, 123, 124, 309
nuclear weapons, 16

O

O'Hanlon, 88, 298
O'Prey, 56, 57, 66
obstacles, 2
ODTC, 174
Office of Defense Trade Controls, 174
offsets, 45, 67
opportunities, 9
Ostry, 38, 184, 283, 299
OTA, 58, 194, 218, 291, 299
Outer Space Treaty, 111, 293
Owens, 178
Ozga, 111, 299

P

Pacific rim, 42
Paige, 101, 299
Pakistan, 17, 18, 23, 25, 26, 31, 35, 36, 40, 47, 53, 64, 69, 72, 73, 74, 75, 78, 79, 80, 84, 85, 86, 87, 88, 89, 100, 115, 116, 119, 125, 130, 189, 194, 260, 261, 263, 287, 288, 302, 305
parochial, 109, 247
participation, 39, 47, 54, 136, 153, 187, 189, 218, 221, 244, 245, 256
partner, 62, 181

patents, 4, 158, 161
patience, 49, 169, 237
payments, 25, 159
Pentagon, 4, 60, 219, 220
per capita, 43
perceptions, 14
Perestroika, 12
Perestroika', 12
performance, 22, 51, 60, 102, 103, 110, 153, 156, 192, 249, 253, 263
Perkovich, 36, 84, 116, 299
Perry, 13, 42, 46, 53, 99, 206, 232, 235, 286, 299, 300
perspectives, xii, 19, 38, 95, 97
pessimist, 12
phases, 148, 162
Pincus, 120, 300
plan, 2
plateau, 69
platform, 6, 31, 42, 111, 143, 258
players, 12
Pokharan, 112
political hurdles, 2
political instability, 41
Pollard, 141, 168, 300
possess, 16
potential conflicts, 11
Powell, 235
Pressler, 25, 53
pressure, 19, 24, 52, 76, 81, 114, 119, 210, 248
prestige, 63, 69, 119
Priest, 94, 300
principles, 32, 43, 161, 199
Prithvi, 24, 25, 26, 114
problems, 14, 15, 36, 40, 50, 58, 61, 63, 64, 65, 69, 70, 72, 73, 80, 97, 98, 99, 104, 105, 128, 135, 166, 168, 170, 179, 182, 193, 196, 197, 211, 212, 227, 240, 250, 253, 262
procurement, 45, 60, 62, 69, 90, 166, 195, 197, 199, 213, 219, 220, 221, 249
procurements, 22, 210
profits, 50, 52, 167, 200
proliferation, 15
Proliferation, 15
proliferation', 15
Proliferation', 15

proposal, 146, 166
provide, 14
Public Law, 144, 152
puppeteer, 52
pyramid, 237

Q

quotation, 146, 249

R

R&D, v, vii, ix, xiii, 7, 21, 38, 45, 55, 58, 64, 89, 101, 131, 138, 151, 153, 154, 155, 156, 166, 177, 178, 180, 182, 183, 184, 189, 190, 192, 194, 195, 196, 197, 198, 199, 201, 203, 206, 210, 227, 238, 248, 252, 254, 263, 264, 266, 271, 303
Raghuvanshi, 83, 90, 187, 300, 301
Raja Mohan, 119, 300
Rajghatta, 213, 301
Rajiv Gandhi, 234
rationale, 5
Ray, 235
Reagan, 134, 234
realism, 37
reallocations, 14
rearming, 93
recipient, 56, 59, 64, 69, 147, 148, 150, 151, 152, 154, 155, 156, 158, 159, 160, 161, 167, 170, 215, 216, 246
Recipient Phase, 150
record, 44, 48, 54, 57, 216, 233, 236, 246, 247
reduction, 15
reductions, 14
Rees-Mogg, 30, 71, 301
regime, 6, 33, 48, 113, 114, 117, 121, 122, 127, 128, 172, 237, 254, 262
Reinicke, 173, 301
Reinsch, 60, 61, 171, 301
relationship, xii, 2, 12, 23, 29, 32, 35, 38, 46, 75, 85, 86, 111, 112, 115, 141, 143, 201, 202, 211, 231, 232, 234, 235, 237, 239, 245, 265, 266
reliance, 45, 125, 126, 131, 143, 168, 172, 178, 181, 185, 186, 187, 188, 189, 194, 207, 238, 249

requirements, xi, 14, 42, 47, 59, 61, 63, 67, 109, 130, 135, 145, 146, 150, 161, 163, 166, 171, 187, 196, 198, 199, 202, 208, 210, 215, 216, 217, 221, 236, 240, 246, 249, 258
research and development, 7, 39, 48, 52, 84, 131, 141, 152, 160, 162, 177, 190, 205, 210, 238, 251, 253, 256, 262
resident, 151
resilience, 42
resilient, 1
resources, 5, 10, 20, 36, 39, 43, 57, 58, 63, 64, 66, 95, 102, 138, 152, 160, 168, 180, 183, 194, 197, 199, 211, 233, 238, 246, 250, 251, 261
responsibility, 14
rethinking, 13
Reuter, 116, 297, 298, 301, 304
Riedel, 238
Risdon, 148, 301
Risk Report, 26, 75, 301
rocket, 20, 195
rockets, 25, 112
Rodan, 47, 301
Rodrigues, 240
rogue nation, 15
rogue nation', 15
Rohini I, 112
roots, 42
Russia, 40, 46, 54, 57, 60, 63, 66, 77, 79, 93, 108, 114, 116, 118, 179, 186, 188, 207, 297
Russian weapons, 45

S

SAARC, 125, 309
sabbaticals, 154, 250
Sadowski, 93, 301
Sagarika, 25
scandal, 50
scars, 43
secondary, 1
second-hand, 108, 187
second-rate, 67
self reliance, xii, 19, 131, 169, 181, 193, 195, 206, 249
self reliant, 7

self-defense, 25, 65, 105, 256
sensitive, 4, 31, 33, 37, 99, 102, 104, 122, 133, 180, 239, 247
Shaker, 118
Shalikashvili, 9, 43, 302
share, 3
shared values, 43
Sharma, 302
sheltered, 6
shoestring, 46
shrinking, 4
Singapore, 34, 69
Singh R.P, 126, 302
SIPRI, 69, 183, 286, 294
slump, 57
Smith, 75, 119, 192, 302, 303
Snow, 243
Software, 97
source, 19, 48, 157
sources, 14, 46, 47, 108, 139, 146, 154, 180, 187, 188, 192, 206, 208, 221, 222, 263
South Asia, 23, 36, 40, 63, 73, 74, 81, 86, 97, 125, 292, 293, 295, 299
South Asian Association for Regional Cooperation, 125
South Korea, 34, 99, 127
Soviet Union, 6, 9, 12, 14, 16, 18, 24, 26, 32, 33, 36, 40, 46, 50, 52, 66, 73, 97, 101, 102, 111, 112, 115, 133, 189, 212, 223
Soviets, 35, 98, 212
Spinney, 184
Spin-off, 162
Sri Lanka, 69, 74, 125
stable, 38, 42, 71, 84, 262, 266
Steinbruner, 208, 211, 286, 303
Stevenson, 142
Stinger, 53
strategic, 2
strategic alliance, 18
strategic consistency, 9
strategic convergence, 33
streamlining, 175, 202
structure, 2, 19, 83, 85, 177, 196, 202, 221, 227, 248, 250, 266, 302
subcontinent, iii, 1, 18, 23, 26, 31, 33, 47, 69, 74, 79, 81, 83, 100, 130, 194, 261

Subrahmanyam, 116, 120, 303
Super Cobra, 54
supercomputers, 94
superpower, 10, 11
supplier, 20, 47, 59, 67, 79, 127, 147, 156, 158, 159, 166, 167, 170, 188, 207, 209, 217, 223, 247

T

tactics', 104
Tahir-Kheli, 99
Taiwan, 53, 67, 99, 127
tanks, 56, 189
Taylor, 51, 68, 304
Technical Barriers, 169
technical trials, 163
Technology Innovation Act, 142
technology push, 139
technology transfer, xi, xii, xiii, xiv, 3, 4, 6, 47, 48, 55, 104, 109, 113, 120, 128, 130, 134, 135, 136, 137, 139, 140, 141, 142, 143, 144, 145, 146, 147, 148, 149, 151, 152, 154, 155, 156, 157, 159, 160, 162, 167, 169, 170, 174, 179, 185, 187, 188, 190, 193, 197, 202, 205, 216, 228, 232, 234, 237, 243, 245, 255, 256, 264
Technology Transfer Act, 142, 152
Technology Transfer Drivers, vii, 139
tenacity, 16, 45, 99
tensions, 16
testing agency, 147, 163
Thailand, 34, 69
the Indian Space Research Organization, 110
third world countries, xii, 6, 20, 21, 48, 142, 219, 264
third-rate, 3
Thoreau, 31
threat, iii, xii, 5, 6, 10, 12, 14, 15, 22, 27, 35, 54, 75, 76, 84, 96, 97, 98, 104, 110, 112, 115, 119, 183, 195, 219, 255, 258
threats, 11
tightrope exercise, 17, 100
time lag, 197
TIUSS, 31, 304
Toffler, 10

trade, 5, 19, 38, 42, 43, 45, 50, 52, 55, 64, 69, 99, 124, 144, 145, 147, 156, 170, 171, 172, 173, 175, 207, 208, 209, 212, 223, 245, 255
trade turnover, 44
trading partner, 43
transfer criteria, 55, 216
Transparency, 232, 293, 295
treaty, 34, 35, 48, 110, 115, 117, 118, 119, 120, 121, 123, 129, 130, 259
Trigger list', 123
Trishul, 228
Truman, 234
trust, 4, 22, 43, 231, 233, 240
turmoil, 9
Turner, 17

U

UN Register, 54, 218
UNISPACE, 228, 304
United States, iv, xiv, 2, 3, 5, 6, 9, 10, 11, 12, 13, 14, 16, 17, 18, 19, 24, 26, 29, 30, 31, 35, 38, 39, 40, 42, 43, 46, 47, 50, 53, 54, 56, 57, 59, 61, 66, 71, 73, 75, 76, 77, 85, 95, 96, 97, 101, 104, 107, 110, 113, 116, 118, 125, 130, 131, 133, 136, 141, 171, 173, 177, 194, 205, 207, 212, 214, 223, 231, 233, 234, 235, 237, 239, 240, 243, 244, 246, 258, 259, 260, 261, 263, 265, 266, 271, 283, 285, 289, 292, 293, 294, 296, 302, 303, 304, 307
Unpredictability, 15
upgrade, 46, 66, 216
Uruguay Round, 43
US Congress, v, 25, 76, 134, 141, 210, 212, 299

V

vendor, 147, 151, 157, 163, 166, 250
Victor, 187, 304
vision, 22, 93, 169, 191

W

Wallop, 49, 304
warriors, 14
warships, 56, 67
Washburn, 214, 304
Washington Post, 27, 94, 287, 291, 292, 297, 300, 302
Wassenaar Arrangement, iv, 79, 121, 122
weapons, 12
Weapons of Mass Destruction, 95, 218, 296, 298, 299
White, 21, 22, 53, 55, 60, 61, 96, 215, 216, 298, 305
White House, 53, 55, 60, 61, 215, 216, 298, 305
Widnall, 10, 12, 97, 179, 206, 305
win, xi, 23, 40, 100, 160, 259
winner, 42
withdrawal, 42, 77
Working Paper, 129
world events, 11
world order, 13, 97, 121, 258
Wright Patterson, 138
WWR, 228, 305
Wydler, 142

Z

Zangger, 123

R. J. Augustus

About The Author

Rayol John Augustus Ph.D.

The author has been twenty five years with the Indian Defense Research and Development Organization with experience in Aerodynamics, Flight Mechanics & Controls and specialization in Software and Flight Simulation. As Adviser (Defense Technology) at the Embassy of India in Washington DC since 1994, he handles all Indian Defense R&D interactions with US Defense Industry and US Government. Dr. Augustus received his Bachelor's (Physics), Bachelor's in Aeronautical Engineering and PG Degree in Management in India; Master's Degree (Aerospace) and Diplome de Specialization in France and Ph.D. (Technology Transfer and International Relations) in USA.

R. J. Augustus

www.ingramcontent.com/pod-product-compliance
Lightning Source LLC
Chambersburg PA
CBHW020627220526
45464CB00001B/50